W0079867

Philosophy of Engineering and Technology

Editor-in-Chief
Pieter E. Vermaas, Department of Philosophy
Delft University of Technology
Delft, The Netherlands

Series Editors

Darryl Cressman, Department of Philosophy
Maastricht University
Maastricht, Limburg, The Netherlands

Neelke Doorn, Faculty of Technology, Policy and Management
Delft University of Technology
Delft, Zuid-Holland, The Netherlands

Editorial Board

Byron Newberry, Baylor University
Waco, TX, USA

Edison Renato Silva, Federal University of Rio de Janeiro
Rio de Janeiro, Rio de Janeiro, Brazil

Editorial Board

Philip Brey, University of Twente
Enschede, Overijssel, The Netherlands

Louis Bucciarelli, School of Engineering
Massachusetts Inst of Tech
Belmont, MA, USA

Michael Davis, Humanities Department
Illinois Institute of Technology
Chicago, IL, USA

Paul Durbin, College of Arts & Sciences
University of Delaware
Newark, DE, USA

Andrew Feenberg, Simon Fraser University
Burnaby, BC, Canada

Luciano Floridi, Department of Philosophy
University of Hertfordshire
Hertfordshire, UK

Jun Fudano, Kanazawa Institute of Technology
Nonoichi, Japan

Sven Ove Hansson, Division of Philosophy
Royal Institute of Technology KTH
Stockholm, Stockholms Län, Sweden

Craig Hanks, Philosophy Department
Texas State University
SAN MARCOS, USA

Vincent F. Hendricks, Center for Information & Bubble Studies
University of Copenhagen
København, Denmark

Don Ihde, Dept of Philosopy
SUNY at Stony brook
Stony Brook, NY, USA

Billy Vaughn Koen, Dept of Mechanical Engineering
University of Texas
Austin, TX, USA

Peter Kroes, Dept. Of Philosophy, TPM
Delft University of Technology
Delft, Zuid-Holland, The Netherlands

Sylvain Lavelle, Ctr for Ethics, Tech & Society
ICAM Paris-Senart Engineering School
Lieusaint-Senart, France

Michael Lynch, Cornell University
ITHACA, NY, USA

Anthonie W. M. Meijers, Dept of Philosophy and Ethics
Eindhoven Univ of Technology
Eindhoven, Noord-Brabant, The Netherlands

Duncan Michael, Ove Arup Foundation
London, UK

Carl Mitcham, Liberal Arts & International Studie
Colorado School of Mines
GOLDEN, CO, USA

Helen Nissenbaum, East Building 7th Fl
New York University
New York, NY, USA

Alfred Nordmann, Institut für Philosophie
Technische Universität Darmstadt
Darmstadt, Hessen, Germany

Joseph C Pitt, Department of Philosophy
Virginia Tech
Blacksburg, VA, USA

Daniel Sarewitz, Consortium for Sci Policy & Outcome
Arizona State University
Washington, DC, USA

Jon Alan Schmidt, Aviation & Federal Group
Burns & McDonnell
Kansas City, MO, USA

Peter Simons, Trinity College Dublin
Dublin, Ireland

Jeroen van den Hoven, Delft University of Technology
Delft, The Netherlands

Ibo van der Poel, Delft University of Technology
Delft, The Netherlands

John Weckert, Centre for Applied Philosophy & Ethics
Charles Sturt University
Canberra, ACT, Australia

The *Philosophy of Engineering and Technology* book series provides the multifaceted and rapidly growing discipline of philosophy of technology with a central overarching and integrative platform.

Specifically it publishes edited volumes and monographs in:

- the phenomenology, anthropology and socio-politics of technology and engineering
- the emergent fields of the ontology and epistemology of artifacts, design, knowledge bases, and instrumentation
- engineering ethics and the ethics of specific technologies ranging from nuclear technologies to the converging nano-, bio-, information and cognitive technologies
- written from philosophical and practitioners' perspectives and authored by philosophers and practitioners

The series also welcomes proposals that bring these fields together or advance philosophy of engineering and technology in other integrative ways.

Proposals should include:

- A short synopsis of the work or the introduction chapter
- The proposed Table of Contents
- The CV of the lead author(s)
- If available: one sample chapter

We aim to make a first decision within 1 month of submission. In case of a positive first decision the work will be provisionally contracted: the final decision about publication will depend upon the result of the anonymous peer review of the complete manuscript. We aim to have the complete work peer-reviewed within 3 months of submission.

The series discourages the submission of manuscripts that contain reprints of previous published material and/or manuscripts that are below 150 pages/75,000 words.

For inquiries and submission of proposals authors can contact the editor-in-chief Pieter Vermaas via: p.e.vermaas@tudelft.nl, or contact one of the associate editors.

Matthew J. Dennis • Georgy Ishmaev
Steven Umbrello • Jeroen van den Hoven

Editors

Values for a Post-Pandemic Future

Springer

Editors
Matthew J. Dennis [iD]
Eindhoven University of Technology
Eindhoven, The Netherlands

Georgy Ishmaev
Delft University of Technology
Delft, The Netherlands

Steven Umbrello [iD]
Delft University of Technology
Delft, The Netherlands

Jeroen van den Hoven [iD]
Delft University of Technology
Delft, The Netherlands

Editor-in-Chief
Pieter E. Vermaas

This book is an open access publication.

ISSN 1879-7202   ISSN 1879-7210   (electronic)
Philosophy of Engineering and Technology
ISBN 978-3-031-08423-2   ISBN 978-3-031-08424-9   (eBook)
https://doi.org/10.1007/978-3-031-08424-9

© The Editor(s) (if applicable) and The Author(s) 2022
Open Access This book is licensed under the terms of the Creative Commons Attribution 4.0
International License (http://creativecommons.org/licenses/by/4.0/), which permits use, sharing,
adaptation, distribution and reproduction in any medium or format, as long as you give appropriate credit
to the original author(s) and the source, provide a link to the Creative Commons license and indicate if
changes were made.
The images or other third party material in this book are included in the book's Creative Commons
license, unless indicated otherwise in a credit line to the material. If material is not included in the book's
Creative Commons license and your intended use is not permitted by statutory regulation or exceeds the
permitted use, you will need to obtain permission directly from the copyright holder.
The use of general descriptive names, registered names, trademarks, service marks, etc. in this publication
does not imply, even in the absence of a specific statement, that such names are exempt from the relevant
protective laws and regulations and therefore free for general use.
The publisher, the authors, and the editors are safe to assume that the advice and information in this book
are believed to be true and accurate at the date of publication. Neither the publisher nor the authors or the
editors give a warranty, expressed or implied, with respect to the material contained herein or for any
errors or omissions that may have been made. The publisher remains neutral with regard to jurisdictional
claims in published maps and institutional affiliations.

This Springer imprint is published by the registered company Springer Nature Switzerland AG
The registered company address is: Gewerbestrasse 11, 6330 Cham, Switzerland

Contents

Contributors

Simisola Akintoye De Montfort University, Leicester, UK

Jan Peter Bergen University of Twente, Enschede, Netherlands

Vincent Blok Department of Communication, Philosophy and Technology, Wageningen University and Research, Wageningen, The Netherlands

Gunter Bombaerts Eindhoven University of Technology, Eindhoven, Netherlands

Jose C. Cañizares-Gaztelu Delft University of Technology, Delft, Netherlands

Samantha Copeland Delft University of Technology, Delft, Netherlands

Alessandro Corbetta Eindhoven University of Technology, Eindhoven, Netherlands

Andrej Dameski Eindhoven University of Technology, Eindhoven, Netherlands

Matthew J. Dennis Eindhoven University of Technology, Eindhoven, The Netherlands

Tristan de Wildt Delft University of Technology, Delft, The Netherlands

Damian Okaibedi Eke De Montfort University, Leicester, UK

Pei-Hua Huang Erasmus Medical Centre, Rotterdam, The Netherlands

Wijnand IJsselsteijn Eindhoven University of Technology, Eindhoven, Netherlands

Georgy Ishmaev Technical University Delft, Delft, The Netherlands

Tonii Leach De Montfort University, Leicester, UK

Kritika Maheshwari University of Groningen, Groningen, Netherlands

Sven Nyholm Utrecht University, Utrecht, Netherlands

Paschal Ochang De Montfort University, Leicester, UK

George Ogoh De Montfort University, Leicester, UK

Adebowale Owoseni De Montfort University, Leicester, UK

Oluyinka Oyeniji De Montfort University, Leicester, UK

Udo Pesch Delft University of Technology, Delft, The Netherlands

Eugen Octav Popa Department of Science Technology and Policy Studies University of Twente, Enschede, The Netherlands

Caspar A. S. Pouw Eindhoven University of Technology, Eindhoven, Netherlands

Zoë Robaey Wageningen University, Wageningen, Netherlands

Sabine Roeser Delft University of Technology, Delft, Netherlands

Andreas Spahn Eindhoven University of Technology, Eindhoven, Netherlands

Bernd Carsten Stahl De Montfort University, Leicester, UK

Federico Toschi Eindhoven University of Technology, Eindhoven, Netherlands

Steven Umbrello Technical University Delft, Delft, The Netherlands

Jeroen van den Hoven Technical University Delft, Delft, The Netherlands

Ibo van de Poel Delft University of Technology, Delft, The Netherlands

Janna van Grunsven Delft University of Technology, Delft, Netherlands

Dyami van Kooten Pássaro Delft University of Technology, Delft, The Netherlands

Elena Ziliotti Delft University of Technology, Delft, The Netherlands

Chapter 1
Values for a Post-Pandemic Future

Matthew J. Dennis ⓘ, Georgy Ishmaev, Steven Umbrello ⓘ, and Jeroen van den Hoven ⓘ

1.1 Value Disruption & COVID-19

At the beginning of the COVID-19 crisis, several public figures predicted that the pandemic would precipitate a dramatic shift towards new sets of values in our societies. Other more optimistic commenters prophesied a new dawn for egalitarian and progressive values (Adib-Moghaddam, 2020; Kelly, 2020; Nancy, 2020). This conjecture was drawn from the early belief that the SARS-CoV-2 virus would be impervious to differences in age, class, ethnicity, and nationhood: a 'great equaliser'. As statistics on death rates and hospitalisation rose, however, this optimism quickly came to be seen as misguided. Not only are some individuals more susceptible to the virus (ethnic minorities, senior citizens, those with pre-existing conditions), the non-medical measures designed to prevent populations from spreading the virus disproportionately affect other marginalised groups (such as those who have less income or education, etc.). When more information became available on the causes, exacerbating factors, and the prognosis of COVID-19 infection, some authorities tried to make medical outcomes more equitable. (1) In some counties, those most at risk from complications from the virus were often (although not always) given earlier treatment or vaccine priority. (2) Some policymakers initially recognised (or at least declared publicly) that disadvantaged communities and critical workers should be vaccinated first. (3) Globally speaking, the World Health Organization's COVAX scheme provided millions of vaccine doses to low-to-middle-income countries.

M. J. Dennis (✉)
Eindhoven University of Technology, Eindhoven, The Netherlands
e-mail: m.j.dennis@tue.nl

G. Ishmaev · S. Umbrello · J. van den Hoven
Technical University Delft, Delft, The Netherlands
e-mail: G.Ishmaev@tudelft.nl; s.umbrello@tudelft.nl; M.J.vandenHoven@tudelft.nl

© The Author(s) 2022
M. J. Dennis et al. (eds.), *Values for a Post-Pandemic Future*, Philosophy of Engineering and Technology 40, https://doi.org/10.1007/978-3-031-08424-9_1

Nevertheless, over two years since the beginning of the pandemic, some of these critical initiatives are still unsuccessful.[1] Vaccine and booster shots are still distributed in a chronically inconsistent manner (Sawal et al., 2021). In international terms, national boundaries have invariably determined which citizens have received jabs first. Furthermore, misinformation regarding vaccine efficacy or full-blown vaccine conspiracy theories has caused a sizeable minority to refuse (even to protest against) having the jab, especially in countries in which governments are actively legislating for this. For example, the WHO's 2021 plea for national governments in the rich world to halt booster shots before the vaccine was rolled out worldwide was entirely ignored (Keaton, 2021). Citizens of the countries without access to internationally recognised vaccines are effectively separated from the first world countries, both in terms of healthcare and travel. Thus it seems that contrary to those who predicted a new sense of social and political connectedness, COVID-19 has fed into further polarisation in societies and made the world even more divided and balkanised. These differences become strikingly apparent in entrenched differences in values: what we think is important, worth preserving, and what we care about as individuals and larger social groups.

While the utopian predictions of the pandemic may look naïve or simple-minded in hindsight, it is also true that new value shifts and conflicts that the pandemic has created were hard to anticipate. Much has been written on cultural differences, for example, specifically the ability of East Asian countries to control the spread of the virus effectively. Mask wearing in these countries is standard, often viewed as a mark of respect for those in one's vicinity. Because of cultural differences, so the argument goes, the governments in some East Asian countries were better able to implement timely lockdowns, strict limits on public transport, and mandatory testing. This has been attributed to various causes. Historical precedents regarding the value trade-off between conformity for social norms and personal liberty may have something to do with it. However, it should also be remembered that many East Asian countries have greater experience with respiratory diseases (SARS, MERS, etc.). By contrast, the initially insouciant reactions of many Western Democratic governments turned into painful and often inconsistent attempts, in the later stages of a pandemic, to balance the COVID-19 containment, economic fallouts, and incursions on individual liberties. The fine-grained picture, of course, is more complicated than this. We know that pandemic containment measures in some Eastern Asian countries were deployed at the cost of brutal suppression of basic citizens' liberties, with some of the Western democratic countries also moving dangerously close in this direction (Greitens, 2020). Contrawise, some East Asian governments, such as South Korea and Taiwan, have become examples of successful deployment of emergency pandemic containment measures without the erosion of democratic processes. We can only hope that the pandemic will serve as a cautionary learning experience for the future. There are some signs of this already. At the time of

---

[1] WHO calls distribution of Covid boosters a 'scandal' as poor nations struggle to get first shots. https://www.cnbc.com/2021/11/12/who-calls-distribution-of-covid-boosters-a-scandal-as-poor-nations-struggle-to-get-first-shots.html

writing, many governments have reacted more rapidly to the threat of the so-called Omicron variant by shutting national borders, as well as enhancing testing and quarantine requirements for those who come from potentially infected regions.

The way populations deal with virus-containment measures has revealed differences in values that may have been implicit before the pandemic but became explicit under lockdown conditions. To take an example from the Benelux region, when the Dutch government mandated the closure of 'non-essential services', they included bookshops in this definition but not florists. In Belgium, by contrast, florists were ordered to close, but bookshops were free to remain open for some time as they were deemed fundamental to the Belgian way of life. Here we can see how policymakers implicitly made value judgements concerning governmental definitions of an 'essential service', deciding what counts as essential to the way of life of the populations they represent. Another upshot has been increasing public discussion of how those who work in essential services are often not remunerated in a way that is commensurate with the importance – *essentialness* – of their role. Compared to those who work in sectors that were easily able to move their workplace from the office to their home (bankers, public servants, accountants, etc.), essential workers (nurses, care workers, bus drivers, teachers, grocery staff, etc.) are often paid significantly less. Whether this disparity will figure in how we value these activities in the future remains to be seen. Will those whose work was defined as 'essential' be remunerated accordingly, or will this definition be forgotten once COVID-19 no longer presents a threat?

The natural environment is another domain that shows signs of being susceptible to post-pandemic value change. In their original explanation of 'building back better', the OECD cautions that 'global environmental emergencies such as climate change and biodiversity loss' are formidable existential threats to humanity (2020: 2). From this, the authors propose, we should view the value disruption of the pandemic as an opportunity to rethink our attitudes to 'long-term emission reduction goals, […] resilience to climate impacts, […] biodiversity loss and […] circularity of supply chains' (2020: 2). We saw intimations of this reappraisal during initial lockdowns as citizens across the world were amazed at the reduction of local smog (Venter et al., 2020), the sharp increase in species population (Natural History Museum, 2020), and beautiful images of Venice canals finally running clean after decades of pollution (Katanich, 2021). However, despite these acute short-term signs, global lockdowns and the net reduction in emissions as a consequence has done little to alleviate the underlying causes of these issues. Perhaps, however, this temporary respite in environmental degradation may raise ecological issues, thinking, and values in the minds of many in a way that will inform how ecological challenges are subsequently approached.

In general, COVID-19 has galvanised a discussion on future values through the initiative to 'build back better', proposed by various governments and global financial institutions (US, UK, EU, World Economic Forum, etc.). This slogan, initially developed by the Organization for Economic Co-operation and Development (OECD), advocates using the disruption caused by the pandemic to create a future world that is more 'equitable', 'sustainable', 'resilient', and one that pays more attention to social 'well-being and inclusiveness' (OECD, 2020: 2; cf. Schwab & Malleret, 2020). How these goals can be achieved, if they are possible at all, remain subject to debate, controversy, and conspiracy (Umbrello, 2021).

1.2   COVID-19 Technologies

Whether the pandemic will change our values in the domains of social justice or sustainability will only be seen in future decades. However, due to the exigencies of the immediate consequences of the pandemic, value issues relating to COVID-19 technologies have come into focus much more rapidly. Ethicists and philosophers of technology usually have significantly longer timeframes to evaluate the impact of emerging technologies. Within weeks of the pandemic emerging, however, various *digital* and *medical* technology companies were vying to show how their products could be repurposed to slow the transmission of the SARS-CoV-2 virus (e.g., COVID-tracking apps, digital immunity passports) or fight the medical effects of COVID-19 (respirators, antigen treatments, mRNA vaccines).

Digital technologies have been promoted as a way to mitigate the indirect social effects of lockdown, the closure of schools and workplaces, and the restrictions on socialising (Alphabet CEO, Eric Schmidt; cited by Strauss, 2020). While schools may have to remain closed, classrooms have been exchanged for online education; while visits to the elderly are banned, video conferencing has replaced family visits; while workplaces are out of bounds, many have worked from home. The problem with digital solutions, as many now recognise, is that they assume some level of socioeconomic parity, as moving one's life to online-only works effectively in a stable and secure home environment. These changes show signs of creating a digital divide that may adversely affect socioeconomically disadvantaged groups. In short, the values and value trade-offs enshrined in a post-COVID future are yet to be understood. Still, the speed and impact of these developments on governments, social institutions, and individual citizens mean that ethical reflection on the values of the post-COVID world is urgently needed. As we explain in the methodology section below, this raises novel challenges for the ethics of technology, requiring responses to the complex value questions raised by the pandemic, often in real-time.

While digital technologies arguably attracted more attention at the early pandemic stages, medical technologies have assumed much more importance for many of us in the last two years. From antigen treatments of the symptoms of the virus to mRNA vaccines, many of us have come into contact with cutting-edge biotechnologies that have had a powerful effect on mitigating the impact of the virus. Permanent funding has been allocated to new virus monitoring stations at national borders. Flu vaccines created with mRNA technology are due to become available as early as 2023. Many of these technologies have been deployed at scale for the first time. These new ways of integrating technology into our everyday lives stand to be one of the legacies of COVID-19 that will have profound ethical consequences.

1.2.1   Contact-Tracing Apps

Throughout the COVID-19 pandemic, several smartphone apps were created to work symbiotically and increase the efficiency of manual contact tracing efforts without knowing beforehand if they would actually be effective. The MIT Technology Review's 'Covid Tracing Tracker' listed around 50 apps globally, 22 of them solely within the European Union. Although some initial data have been produced, and these preliminary findings can and should be explored and debated, the evidence concerning the real-world effectiveness of these digital contact tracing apps remains both unclear and, in many cases, contradictory (Tupper et al., 2021; Keeling et al., 2020).

When the pandemic began, the idea behind digital contact and tracking apps was to 'solve' the pandemic itself. However, we know now that framing these apps within a 'technofix' picture is not ideal for conceptualising their use. The data we have concerning their effectiveness diverge significantly from country to country and, at times, from study to study. The variation in the methods employed are so divergent that it is difficult – if not entirely impossible – to compare results and come up with a coherent, comprehensive evaluation of the impact and effectiveness of digital contact tracing apps in actual responses to the COVID-19 pandemic. And despite the revealed privacy abuses of many of these apps, governments are now downplaying their relative importance, a drastic change in value priorities compared to their initial interest and investment early on in the pandemic.

Regardless, in the end, even the uptake of these apps was not sufficient to meet the minimum necessary number of downloads to be considered effective (c.f., Sabelli, 2021). As of July of 2021, 17% of Italy's population downloaded such apps, Spain at 16%, Poland at 4%, and Croatia at 2% (LibertiesEU, 2021). What was at one time, not far in the past, of value has become ever more hushed.

1.2.2   Immunity Passports

Another technology that has taken up a great deal of real-estate on the front page news is immunity passports or digital covid certificates. Created to aid in the reopening of international travel, digital immunity passports are now mandatory in several countries to enter premises such as bars, restaurants, gyms, pools, and museums and attend large public events. In fact, Italy, one of the countries that mandated such passports (often termed the 'green pass'), created a 'super green pass' of sorts that prohibited people from attending school or work if they didn't have one (Italian Committee for Bioethics, 2021).

If we look at the available literature on the use of immunity passports and historical precedents, some evidence and arguments indicate that there are shared concerns within the scientific community that are being overlooked or downplayed by governments that have adopted immunity passports for domestic use (Milan et al.,

2021). This fact has become even more critical given that the debate – or lack thereof – concerning immunity passports fits into the broader landscape of emergency technological responses to the COVID-19 pandemic. Unfortunately, these measures are often tainted by controversy pertaining to their lack of transparency, evidence of efficiency, similar to the accusations and critiques levelled at digital contact and tracing apps. Consequently, an increasing number of studies suggest that the actual contribution of immunity passports in combating COVID-19 – both in terms of boosting vaccination rates and containing infections – could be more controversial than how governments who uncritically pushed for their adoption would like it to be. Hastily deploying them could lead to increased polarisation concerning vaccine hesitancy and rejection from certain demographic groups of the population while only achieving marginal results among those who comply with the program (de Figueiredo et al., 2021; Porat et al., 2021).

Governments across the globe appear to have forgotten about contact tracing apps, shifting their communicative strategies to manufacture consent for immunity passports. The broad use of narratives offering a different rationale for these programs is also of particular concern here. From the arguments that imposing restrictions on vaccinated individuals is unfair to the arguments that COVID-19 passports can avoid hard lockdowns to the open admissions that these programs are devised to nudge more people into vaccination. It is challenging to ignore parallels with the deployment of contact-tracing apps here. What is becoming more apparent is that domestic immunity passport programs fit into the same trend as contact-tracing apps: "technofixes" to the pandemic, deployed at the expense of the normalisation of health surveillance devices (Kravchenko & Karpova, 2020). However, in contrast to contact tracing apps, an individual cannot choose not to use an immunity passport as COVID-19 passport programs are de-facto obligatory for everyday activities in many countries.

Concerning both technologies, governments are focused on the production of narratives about effectiveness and desirability of these technologies. They are doing so to gain public adoption and participation. This performance fabricates the impression of efficacy on the government's part while repressing critique and resistance.[2] What is fundamentally required is precisely not this, but open and transparent evidence-based dialogue given the constantly developing scientific knowledge regarding the epidemiological processes involved in COVID-19 transmission.

---

[2] At the time of writing (December 2021), EU countries with COVID passport programs are resorting to closing borders and hard lockdowns again, suggesting that the only justification that has not been refuted empirically is the nudge towards higher vaccination uptake.

1.2.3 Novel Antivirals and Vaccines

Aside from the multiple vaccines available for prophylactic use against COVID-19, there are extant treatments with varying levels of efficiency in combating the virus, including anti-inflammatories like dexamethasone and tocilizumab, antivirals like ivermectin, and monoclonal antibodies like casirivimab and imdevimab (Ajayi, 2021; Alam et al., 2021; Francés-Monerris et al., 2021; Mody et al., 2021). Pfizer has recently announced a novel antiviral drug designed to treat COVID-19. This novel drug has been shown to be highly efficacious in preventing severe disease and hospitalisation (Pfizer, 2021). However, pharmacodynamic analysis reveals that the modality of action of this novel drug is similar to that of the generic extent drug Ivermectin (c.f., Francés-Monerris et al., 2021; Mody et al., 2021). This contradiction has produced some polarised debates on the choice of different treatments both within and outside of scientific community (Izcovich et al., 2021). On one hand, preliminary studies suggest that generic, non-patented existent treatments, like Ivermectin, show the highest binding affinity with the virus spike protein (see Eweas et al., 2021; Francés-Monerris et al., 2021; see also Surti et al., 2020; Mody et al., 2021). On the other hand, given the example of Ivermectin, closer scrutiny of meta-analysis studies claiming benefits of this treatment suggests some quality issues (Lawrence et al., 2021; Izcovich et al., 2021).[3] And as Lawrence et al. (2021) argue we should not ignore severe harms and moral hazards that lack of proper scrutiny for the quality of scientific research can bring in the context of unfolding pandemic. This caution, however, should not obscure a valid concern that patents can be highly profitable, which is often understood as motivating the creation of new drugs, despite existing drugs offering similar efficacy.

The open and transparent debate on the efficacy and affordability of different COVID-19 treatments, unfortunately, has been obscured by the extreme politicisation of these topics. Many Western media outlets have campaigned to politicise the use of such drugs, turning the scientific research on safety and effectiveness into supporting arguments for polarised political debates (Szawarski & Rich, 2021). Like the discourse on contact and tracing apps, such media conglomerates have made determined efforts to manufacture consent for the use of patented vaccines and drugs at the opportunity cost of their off-patent counterparts. The reasons behind this push are not difficult to understand given the size of the pharmaceutical lobby in pressuring Western governments as well as their open and costly campaign in advertising their products on widely disseminated media outlets (Merelli, 2021). These efforts are not dissimilar to disturbing lobbying efforts by commercial

---

[3] First large-scale randomised trial on the efficacy of Ivermectin for COVID-19 treatment that will provide conclusive evidence is still underway at the moment of the writing (December 2021). https://www.ox.ac.uk/news/2021-06-23-ivermectin-be-investigated-possible-treatment-covid-19-oxford-s-principle-trial

companies who would like to deploy permanent digital identity solutions piggy-backing on 'COVID-19 passports'.[4]

Commercial interests of Big Pharma now seem to be deeply intertwined with the introduction of 'COVID-19 passports' and other initiatives making vaccination de-facto obligatory. This context does warrant certain scepticism and suspicion about statements from these companies about high desirability of booster shots, feeding into proposals to accelerate 'booster' vaccinations, proposing third, fourth, and more booster doses. This accelerated demand for booster shots is concentrated in richer countries, while populations in developing countries have no access to first doses of efficient vaccines and treatments. All these observations remind us that the long-term effects of these varied corporate interests are slowly becoming manifest, and we must be cognisant of the damages that will emerge in the future.

1.3 Methodological Issues

The unfolding crisis of the COVID-19 pandemic has uprooted value hierarchies in societies across the world. It has also warped our perception of space and times, often in the most peculiar and unexpected ways. Not only has the pandemic demon-strated how globalised our world is, but it has highlighted the increased pace of many technological developments on the global scale. Research and development lifecycles have been accelerated dramatically, bringing spectacular scientific break-throughs such as new COVID treatments, as well as complex challenges. These shortened lifecycles mean that sometimes raw technological solutions were deployed at scale without proper assessments of safety, security and ethical issues (Ishmaev et al., 2021; Lanzing, 2021). Furthermore, these examples have made it evident that these large scale and high impact deployments are at the liberty of a handful of gatekeepers, like pharmaceutical giants or digital platforms.

These challenges pose hard questions to the ethics of technology. For one, the traditional methods of conceptual reasoning based on established academic publica-tions fail to keep up with these developments. Secondly, academic research on the ethics of emerging technologies traditionally operates with a certain degree of detachment, focusing on potential issues in the future or issues that may be relevant only to a small number of people acting as early adopters of new tech. However, research challenges brought by pandemic technologies turned out to be very differ-ent, characterised by unprecedented empirical complexity and moral weight.

The complexity factor has brought a critical value of cross-field communication to the forefront, raising the bar for minimally meaningful contributions from ethics research. This means that in the same way technologists and governments could be accused of 'techno-fixes' and silver-bullet thinking; ethicists could be charged

---

[4]Wetenschappers waarschuwen voor een nieuwe digitale identiteit. https://www.ftm.nl/artikelen/internationale-digid-lobby

with naive 'black-box' technology evaluation methodologies. Contact-tracing apps are just one example that highlighted the need for systemic assessments, combining both high-level societal thinking and empirically grounded low-level evaluation of technical components (Klenk & Duijf, 2021). From the low-level point of view, as it turned out, even seemingly obscure details regarding limitations of a Bluetooth protocol or choice of encryption schemes made a dramatic difference between somewhat practical and secure applications and completely useless solutions ripe with ethical issues (Troncoso, 2021). But these examples have also highlighted that even the most technologically sound solutions can be ethically problematic in ways that they get embedded in other systems and structures of our society (Sharon, 2021).

The moral weight of these ethical issues has also put to the test the value of abstract conceptual reasoning when dealing with urgent and impactful issues. Deployment of many technological solutions in this crisis was characterised by the hard path dependences, such as politics, commercial interests, and even ideologies. This made it much harder for the ethicists to enjoy the ivory tower detachment of moral-theoretical realms. It made even the most well-meaning moral reasoning on the acceptability or desirability of new technologies precariously vulnerable to the co-option by unscrupulous parties. The phenomenon of 'ethics-washing', where public or private actors selectively shop for ethical principles most fitting their practices, emerged before the pandemic. But in the course of the pandemic years, it got entangled with the 'COVID-washing' of questionable technologies and complex institutional arrangements (Ishmaev et al., 2021). This has created an uneasy background where an ethical analysis on the pandemic technologies can be easily co-opted, for example, by the proponents of the radical anti-vaccination movement or proponents of intrusive digital surveillance.

All these challenges have made it clear that, like never before, ethicists have to exercise great epistemic humility without shirking the moral responsibility of expert judgments on the issues of moral import. However, these challenges also provide unique opportunities to advance the field of the ethics of technology. These accelerated innovations present an invaluable opportunity to study full lifecycles of technological solutions from speculative proposals to mass-scale adoption in a span of a few months. Ethicists are presented with invaluable case studies that provide insights on how speculative technologies succeed or fail, their hard path dependencies, and the value conflicts they provoke. It is also an opportunity for the ethicists to reflect on their respective fields' methodologies and research goals. This edited volume presents a step in this direction. It brings about various types of ethical investigations dealing with some of the most challenging topics of the moment, from narrow applied issues to meta reflections on the role of academic ethics research in the crisis.

1.4   Values for a Post-Pandemic Future

Thinking about post-COVID values requires comparing *what shows signs of permanently changing* with *what is likely to stay the same* after the current pandemic has passed. It also requires us to acknowledge that a desirable post-pandemic future can only be better prepared for rather than fully achieved. Although the last two years have shaken the axiological assumptions of many, the values of the post-COVID world will be profoundly influenced by how we have collectively had to reorganise our lives during the last two years. Extended lockdowns have required a collective rethinking of how we work, shop, study, entertain ourselves, and care for each other. In the months and years ahead, we will see whether these new practices have translated into a wholesale re-valuation of the values we live by or a reappraisal of our obligations and duties to one another. Whether these intentions are shelved once the pandemic is over remains to be seen, of course, but there are at least signs that they may well have some longevity, not the least due to the seismic economic disruption the pandemic has caused.

The questions we face then are how to confront and live *with* these new frameworks and baselines of 'normality' and whether or not we *should* live with the changes that have and will be pervasive in a post-pandemic world. Despite many things changing, many pre-pandemic issues have nonetheless remained or have been exacerbated. The Pareto Principle has reared its ugly head in its most devastating form not long after the United States found itself in the grips of the pandemic. Despite a national health emergency putting federal, state, and local resources and infrastructures to the test, the US Congress was quick to put the CARES Act into place, a seemingly necessary piece of legislation that amounted to nothing other than the most significant upward transfer of wealth in the history of humankind (Gross, 2020). Amid a pandemic, the already wealthy class looted the treasury at the expense of those who already had nothing. Some things change, some things stay the same (Abramson, 2020).

When considering the values of the post-pandemic future, we need not only consider *what* we value and *when*, but the *how* of values, i.e., valuation. From the onset of the pandemic to today, medical staff around the globe have rightfully been raised to the station of 'heroes' given the gruelling conditions and constant threat of danger that they constantly confront and continue to face daily. Although their work has always faced danger, and their valuation as heroes *should* have always been such, the unique pandemic crises made manifest and exacerbated what has always been there. But the tides are changing for our heroes. Confronted with mandatory vaccine mandates, large swaths of medical workers are resigning from their posts, suspended without pay, or dismissed entirely from their positions (Kelly, 2021). These 'heroes' are now being lumped in with anti-vax radicals and condemned to stigma and unemployment in a situation that requires their expertise. *How* has this change in what we value come about so quickly?

This dynamism between static, exacerbated, and changing values in light of this global pandemic is the issue to which this curated volume is dedicated.

1.5   Overview of Contributions

As the title of this volume suggests, there is increasing interest in the uncertain future we are moving into, given the current pandemic situation. The values that will come to be held dear, given the continually dynamic and changing nature of our politics, technology, and society as a whole, will have inextricable impacts on our day-to-day lives and how we understand our place in this changing world. To confront these challenging issues, the contributions of this volume have been divided up into two thematic parts. In *Part I: Learning from COVID-19*, the chapters explore the invaluable experiences, values, changes, and issues that have emerged as a consequence of the pandemic situations. Composed of seven chapters, Part I aims to provide us with a solid background to guide us to understand better what our future may hold. We cannot know where we are going if we do not first learn from where we have come from. *Part II: Envisioning a Post-Pandemic Future* takes up these foundational lessons and casts our minds into what our future post-COVID may look like, given historical and current trends. Composed of five chapters, Part II will guide the interested reader along a series of possible futures concerning how we understand the 'new normal', how we can educate the innovators of tomorrow with the lessons of today, as well as how we can guide our behaviour towards socially beneficial ends. Taken together, the two parts aim to give the reader a detailed roadmap to navigate this tenuous and precarious landscape.

Ibo van de Poel, Tristan de Wildt, and Dyami van Kooten Pássaro begin our guided tour of this landscape in their chapter *COVID-19 and Changing Values*. Their chapter takes a close look at value change due to the corona pandemic. With the help of topic modelling, they analysed COVID-related news articles for changes in the frequency of how often these news articles address eleven different values. They found that in the first few months of the pandemic, there was a punctuated shock in the frequency in which values were addressed. They highlighted a sharp increase in the value of health and safety and a significant decline in the values of democracy, privacy and socio-economic equality. However, they noted an opposite direction of change after the first months, which suggests that the punctuated shock's effect may be cancelled over time. Their chapter also presents – and offers possible explanations for – differences between countries and compares their results with the literature. They do not find evidence that the corona pandemic confronts us with a moral dilemma of health versus economic welfare, or lives versus livelihoods, as has sometimes been suggested. Their study also indicates a degree of moral resilience in the studied countries, in the sense of paying attention to morally important values despite being put under pressure during a crisis.

Elena Ziliotti follows van de Poel's analysis in her chapter *What Has COVID-19 Taught Us About Democracy? Relational Democracy and Digital Surveillance Technologies* asking what is the best way for democratic societies to experiment with digital surveillance technologies. This chapter contributes to answering this question through the analysis of the relational democratic model. Ziliotti contends that the relational conception of democracy is a viable approach to experimentations with new technologies. She argues that the relational conception of democracy, which views democracy as a way of life (or culture), supports a deliberative and context-sensitive approach to new digital technologies. To clarify what this approach entails in practice, the chapter discusses the case of South Korea's introduction of new digital surveillance technologies during the first year of the COVID-19 pandemic. Ziliotti demonstrates that these reflections shed new light on what democracy means and provide us with valuable insights on designing post-pandemic democracies.

In their chapter *Contact Tracing Apps for the COVID Pandemic: a Responsible Innovation Perspective,* George Ogoh et alia explore how the COVID-19 pandemic has brought about the first real opportunity to test the efficacy of the Responsible Research and Innovation framework (RRI) in a global health crisis. This is in view of the bold new approaches to health research and innovation that the pandemic has paved the way for. One such approach is the digital contact tracing application (CTA). Although contact tracing has been a fundamental part of infectious disease control for decades, this is the first time this technique has been used in mobile applications. Based on a Multivocal Literature Review, the development of CTAs in four countries – France, Germany, Spain, and the UK – is assessed in this chapter to understand what dimensions of RRI can be identified in the governments' response to COVID-19. This chapter shows that although from 2011, RRI has been promoted as a governance approach for increasing societal desirability of the processes and products of science and technology, very little is known about how the framework may be applied in a health crisis. Ogoh and company show that while no RRI approach was explicitly embraced by these governments, some key components were present - even though inadequately. They argue that this indicates that while it is challenging to apply RRI in crises, there is value in using it as an analytical tool for techno-social responses in situations like those created by the COVID-19 health crisis.

Where Ogoh et alia took up the topic of contact tracing apps, Pei-Hua Huang takes a closer look at vaccines and state duty in her chapter *Uncertainty, Vaccination, and the Duties of Liberal States.* She points out that while a liberal state has a general duty to protect its people from undue health risks, the unprecedented emergent measures against the COVID-19 pandemic give rise to questions regarding the extent to which this duty may be used to justify the intervention. In this chapter, Huang uses the case of vaccination to argue that while a liberal state has a general duty to protect its people's health, the duty cannot be used to justify all sorts of measures. First, every available option involves different risks and benefits. The incommensurability of the involved risks and benefits forbid the prioritisation of a particular vaccine. Second, given the epistemic limitations and uncertainty, policies that favour certain vaccines are not only epistemically ill-founded but also morally problematic. She concludes that in a highly uncertain situation, the duty a liberal state ought to uphold is to properly communicate the knowns and the unknowns to the general public and help people decide which option they'd opt for. Huang calls this duty 'the duty to facilitate risk-taking'.

Eugen Popa and Vincent Blok turn our guide towards the role and impacts of conspiracism in RRI in their chapter *Conspiracism as a litmus test for responsible innovation*. The inclusion of publics in the innovation process has always been the creed of Responsible Research and Innovation (RRI), Public Engagement with Science (PES) and other related fields. Conspiracists, however, are not your garden-variety public. As the COVID-19 pandemic has shown, the conflict between conspiracists and science is deep and intractable – distrust replaces trust, and alternative explanations replace the mainstream narrative. In their chapter, Popa and Blok ask how the game of responsible research and innovation is to be played with those who believe that the game of research and innovation is rigged. Understanding the relationship between conspiracism and responsible innovation is necessary to understand the unvisited corners of the science-society interface in the post-pandemic future. They claim that pluralism, already part of the philosophical background that spurred RRI and PES, can offer insights into how conspiracism can be approached. As a case in point, the authors develop these insights starting from the 2021 E.U. Commission policy on how institutions should respond to conspiracism. They conclude that only within a pluralist framework can RRI and PES become what Sheila Jasanoff referred to as 'technologies of humility'. They conclude by summarizing the distinction between monism and pluralism and by highlighting the consequences of this distinction for concept of 'inclusion' in responsible innovation.

To conclude the first part of this volume on lessons learned, Udo Pesch begins our journey of looking forward. In his chapter *Values as Hypotheses and Messy Institutions: What Ethics Can Learn From the COVID-19 Crisis*, Pesch frames the COVID-19 crisis as an episode that reveals various complications in the relation between values and institutions. He argues that these complications cannot be addressed satisfactorily by ethics, as this field is characterised by a gap between the identification of values worth pursuing and the effectuation of these values in society through politics. His chapter aims to bridge this gap between ethics and politics by outlining the dialectical relation between values and institutions. He does this by firstly presenting values as collectively held understandings that emerge in public deliberation. Secondly, these values are safeguarded by setting up appropriate institutions, which, at the same time, also allows the further substantiation of these values. However, it also needs to be acknowledged that institutions are not mere instrumental solutions to further societal values. On the contrary, they have their own morally laden dynamics. As such, they should also be susceptible to adjustment following societal demand.

In envisioning our potential post-pandemic future, Sven Nyholm and Kritika Maheshwari begin our explorations in their chapter *Offsetting Present Risks, Preempting Future Harms, and Transitioning Towards a 'New Normal'*. The ongoing pandemic has led some people to speak about a 'new normal', since we have temporarily had to radically change how we live our lives to protect ourselves and others from the spread of the SARS-CoV-2 virus. However, the expression – 'a new normal' – has also been used in other contexts, such as in relation to societal disruptions brought about by things like new technologies or climate change. What this general idea of a 'new normal' means is unclear and hard to characterise, and there are diverging views about how to respond to a new normal. Still, one feature of a desirable new normal that most people would agree on is that it should be 'safer': safer technologies, safer institutions, and so on. But it is also essential to consider what other ethical considerations and principles should be part of an ethics of a new normal. And it is also interesting to explore similarities and differences among different types of cases that can be classified as situations where we face a new normal. In this chapter, Nyholm and Maheshwari discuss the general idea of an ethics of a new normal and consider what ethical distinctions, values, and principles are likely to be relevant in most instances where we face a new normal, including ethical considerations related to risk mitigation and ways of offsetting potential harms.

Making this new normal a reality means educating the innovators of the future. In their chapter, *Designing in Times of Uncertainty: What Virtue Ethics Can Bring to Engineering Ethics in the 21st Century?* Jan Peter Bergen and Zoë Robaey take a closer look at the renewed interest in virtue ethics within the ethics of technology scholarship. In their chapter, they explore what virtue ethics can bring to engineering ethics in these times of growing epistemic and normative uncertainty, i.e., when fully informed design choices and trade-offs become increasingly difficult to make. Bergen and Robaey argue that virtue ethics can help us 'do the right thing, at the right moment' in the context of engineering design in different situations of uncertainty.

The COVID-19 pandemic has brought about a pervasive digitalization of our social and practical lives. For many, this has signified a substantial loss, with the pandemic underscoring that in-person interactions play a key if not constitutive role in well-being. At the same time, many disabled people and disability rights activists have celebrated the increased accessibility to practical and social spaces enabled by the pandemic-induced embracing of online communication platforms and other digital technologies. With that, the pandemic offers the opportunity to explore the meaning and value of accessibility and what it means for accessibility to be promoted through technological interventions. This exploration is offered by Janna van Grunsven and Wijnand IJsselsteijn in their chapter, *Confronting Ableism in a Post-COVID World: Designing for World-Familiarity Through Acts of Defamiliarization*. Van Grunsven and IJsselsteijn argue that promoting accessibility involves a readiness to oscillate between two normative imperatives: (1) recognising how human well-being depends on what they term 'world-familiarity,' which can be promoted or thwarted through design and (2) recognizing how world-familiarity can harbour pernicious ableist biases that can be called into question through material gestures of defamiliarization. By presenting these two perspectives as mutually required in the design for accessibility, Van Grunsven and IJsselsteijn hope to better enable technologists and laypersons alike to reflectively evaluate if and how a technological innovation may (or may not) be access-promoting, such that it can contribute to a more just post-COVID world.

In the chapter, *Understanding Risks and Moral Emotions in the Context of COVID-19 Policy Making*, Sabine Roeser looks at how the COVID-19 pandemic crisis highlights how the understanding of and decision making about risk always requires intrinsically ethical considerations in addition to scientific knowledge. Roeser argues that we need to consider the insights of virologists and medical experts, but we also need expertise from ethicists, social scientists, and practitioners in the arts and humanities, as well as involving the public in deliberation. Moral emotions can help bring social and ethical considerations into focus, especially in our collective evaluation of risk. Her chapter argues that moral emotions must be harnessed when designing policies to deal with pandemics: in addition to safety measures, our rich human capacities must inform such policies.

Parallel to the role of moral emotions in design, technologies also shape and guide our behaviours. In their chapter, *How to Balance Individual and Collective Values After COVID-19? Ethical Reflections on Crowd Management at Dutch Train Stations,* Andrej Dameski, Andreas Spahn, and Gunter Bombaerts explore the shift in the balance of individual versus collective values that were instigated by the COVID-19 pandemic. The incredible viral spread rate among the population and its relatively high fatality rate has initially resulted in an assertion of the importance of collective values (such as safety, collective responsibility, and conformism). In contrast, individual rights and values (such as autonomy, freedom, individual responsibility, and privacy) took a 'back seat' for the good of the collective. However, as the pandemic extended over months, there was pressure to reject the primacy of collective values and restore individual values' importance. For example, suppose we wish to return to a healthy and prosperous living within a well-functioning society. In that case, this balance shift between collective and individual values will have to be re-negotiated and resolved to a socially acceptable balance position. The authors undertake this ethical exploration through the lenses of recent changes in how particular technologies were used before and during the COVID-19 pandemic. More precisely, the authors identify and explore broad trends we see relevant to ethics, such as crowd nudging, privacy violations, as well as personal and crowd tracking, with a particular focus on crowd management and the balance shift between individual and collective values as well as individual and collective responsibility.

Samantha Copeland and Jose Cañizares Gaztelu take us to the end of our guided journey with an exploration of narratives and their importance in the creation of our coronial futures. In their chapter, *Rhetorics of Resilience and Extended Crises: Reasoning in the Moral Situation of Our Post-Pandemic World,* Copeland and Gaztelu look closely at the impact of the intersection of the ethics of personal, society and global resilience by first describing the levels of resilience rhetoric at play in the media we use to assess both our own and the situations of our loved ones from afar while we are in lock-down. The authors highlight the conjuncts and disjuncts that can shape our perception both of the resilience and also of the morality of the society we or others are surviving within, more locally speaking. That is, the intersection of personal and global resilience at the level of the community has led to an overlap of concerns, resulting for example in judgments made about local behaviour but based on global experiences. The authors conclude their chapter by looking at how this lock-down experience may have a longer term impact on how we conceive of resilience and its relation to ethics.

Acknowledgements Edited volumes are collective projects, so credit needs to be shared by multiple contributors. First, we would like to thank the authors of the chapters for their incredible contributions, as well as for their timely responses to our reviewers. Second, we would like to thank Pieter Vermaas, the series editor, for his encouraging words on our volume proposal in December 2020. Third, we would like to thank Christopher Coughlin at Springer's New York office for his publishing expertise and assistance with the open access contract. We would also like to thank those who contributed to the research activities upon which this volume. These include the authors of a special issue of *Ethics & Information Technology* (2020), the organisers of the 4TU. Ethics COVID-19 podcast, and the researchers of the 4TU.Ethics & Delft Design for Values COVID-19 Working Group, which began discussing the ethics of COVID-19 technologies as early as April 2020. Finally, we would like to thank Marie Skłodowska-Curie Actions (grant number 707404) and Delft Design for Values for providing the funding to ensure this volume can be read free of charge.

References

Abramson, A. (2020, June 18). Why the trillion-dollar bailout benefited the rich. *Time*. Retrieved November 20, 2021, from https://time.com/5845116/coronavirus-bailout-rich-richer/

Adib-Moghaddam, A. (2020). *A 13th-century Persian poem shows why humanity needs a global response to COVID-19.* Accessed on 11th May 2021. https://thewire.in/culture/13th-century-persian-poem-humanity-coronavirus

Ajayi, A. A. (2021). Drugs shown to inhibit SARS-CoV-2 in COVID-19 disease: Comparative basic and clinical pharmacology of Molnupiravir and Ivermectin. *Austin Journal of Pharmacology and Therapeutics, 9*(5), 1149.

Alam, S., Kamal, T. B., Sarker, M. M. R., Zhou, J. R., Rahman, S. A., & Mohamed, I. N. (2021). Therapeutic effectiveness and safety of repurposing drugs for the treatment of COVID-19: Position standing in 2021. *Frontiers in Pharmacology, 12*, 659577. https://doi.org/10.3389/fphar.2021.659577

de Figueiredo, A., Larson, H. J., & Reicher, S. D. (2021). The potential impact of vaccine passports on inclination to accept COVID-19 vaccinations in the United Kingdom: Evidence from a large cross-sectional survey and modelling study. *EClinicalMedicine, 40*, 101109. https://doi.org/10.1016/j.eclinm.2021.101109

Eweas, A. F., Alhossary, A. A., & Abdel-Moneim, A. S. (2021). Molecular docking reveals Ivermectin and Remdesivir as potential repurposed drugs against SARS-CoV-2. *Frontiers in Microbiology, 11*, 3602. https://doi.org/10.3389/fmicb.2020.592908

Francés-Monerris, A., García-Iriepa, C., Iriepa, I., Hognon, C., Miclot, T., Barone, G., ... Marazzi, M. (2021). Microscopic interactions between ivermectin and key human and viral proteins involved in SARS-CoV-2 infection. *Physical Chemistry Chemical Physics, 23*(40), 22957–22971. https://doi.org/10.1039/D1CP02967C

Greitens, S. (2020). Surveillance, security, and liberal democracy in the post-COVID world. *International Organization, 74*(S1), E169–E190. https://doi.org/10.1017/S0020818320000417

Gross, T. (2020, April 30). How the cares act became a tax-break Bonanza for the rich, explained. *NPR*. Retrieved November 20, 2021, from https://www.npr.org/2020/04/30/848321204/how-the-cares-act-became-a-tax-break-bonanza-for-the-rich-explained

Ishmaev, G., Dennis, M., & van den Hoven, M. J. (2021). Ethics in the COVID-19 pandemic: Myths, false dilemmas, and moral overload. *Ethics and Information Technology, 23*(S1), 19–34. https://doi.org/10.1007/s10676-020-09568-6

Italian Committee for Bioethics. (2021). *Vaccine passport, certificate and green pass, within the Covid-19 pandemic: Bioethical aspects*, 1–10. Rome. Retrieved November 20, 2021, from https://bioetica.governo.it/en/opinions/opinions-responses/vaccine-passport-certificate-and-green-pass-within-the-covid-19-pandemic-bioethical-aspects/

Izcovich, A., Peiris, S., Ragusa, M., Tortosa, F., Rada, G., Aldighieri, S., & Reveiz, L. (2021). Bias as a source of inconsistency in ivermectin trials for COVID-19: A systematic review. Ivermectin's suggested benefits are mainly based on potentially biased results. *Journal of Clinical Epidemiology, 144*, 43–55. https://doi.org/10.1016/j.jclinepi.2021.12.018

Katanich, D. (2021, January 25). Have the canals in Venice really benefited from the lock-down? *Euronews*. Retrieved November 20, 2021, from https://www.euronews.com/green/2020/05/07/what-is-venice-s-real-ecological-profit-from-the-lockdown

Keaton, J. (2021). Who chief urges halt to booster shots for rest of the year. The Associate Press. https://apnews.com/article/business-healthcoronavirus-pandemic-united-nations-world-health-organization-6384ff91c399679824311ac26e3c768a

Keeling, M. J., Hollingsworth, T. D., & Read, J. M. (2020). Efficacy of contact tracing for the containment of the 2019 novel coronavirus (COVID-19). *Journal of Epidemiology and Community Health, 74*(10), 861–866. https://doi.org/10.1136/jech-2020-214051

Kelly, J. (2021, October 1). In a dramatic turn, the once-heralded nurses and healthcare workers are being fired for not getting their vaccination shots. *Forbes*. Retrieved November 20, 2021, from https://www.forbes.com/sites/jackkelly/2021/09/30/in-a-dramatic-turn-the-once-herald-nurses-and-healthcare-workers-are-being-fired-for-not-getting-their-vaccination-shots/?sh=345b25602b62

Kelly, S. (2020, March 21). I spent a year in space, and I have tips on isolation to share. *The New York Times*. Retrieved November 20, 2021, from https://www.nytimes.com/2020/03/21/opinion/scott-kelly-coronavirus-isolation.html

Klenk, M., & Duijf, H. (2021). Ethics of digital contact tracing and COVID-19: Who is (not) free to go? *Ethics and Information Technology, 23*(S1), 69–77. https://doi.org/10.1007/s10676-020-09544-0

Kravchenko, S. A., & Karpova, D. N. (2020). The rationalisation of the surveillance: From the 'Society of the Digital Society and beyond. *Montenegrin Journal of Economics, 16*(3), 197–206. https://doi.org/10.14254/1800-5845/2020.16-3.16

Lanzing, M. (2021). Contact tracing apps: An ethical roadmap. *Ethics and Information Technology, 23*(S1), 87–90. https://doi.org/10.1007/s10676-020-09548-w

Lawrence, J. M., Meyerowitz-Katz, G., Heathers, J. A., Brown, N. J., & Sheldrick, K. A. (2021). The lesson of ivermectin: Meta-analyses based on summary data alone are inherently unreliable. *Nature Medicine, 27*(11), 1853–1854. https://doi.org/10.1038/s41591-021-01535-y

LibertiesEU. (2021, June 2). *Covid-19 contact tracing apps in the EU*. Liberties.eu. Retrieved November 20, 2021, from https://www.liberties.eu/en/stories/trackerhub1-mainpage/43437

Merelli, A. (2021, October 1). Why is pfizer advertising a vaccine that gets plenty of free promotion? *Quartz*. Retrieved November 20, 2021, from https://qz.com/2059769/pfizer-is-planning-to-advertise-its-covid-19-vaccine-comirnaty/

Milan, S., Veale, M., Taylor, L., & Gürses, S. (2021). Promises made to be broken: Performance and performativity in digital vaccine and immunity certification. *European Journal of Risk Regulation, 12*(2), 382–392. https://doi.org/10.1017/err.2021.26

Mody, V., Ho, J., Wills, S., Mawri, A., Lawson, L., Ebert, M. C., … Taval, S. (2021). Identification of 3-chymotrypsin like protease (3CLPro) inhibitors as potential anti-SARS-CoV-2 agents. *Communications Biology, 4*(1), 1–10. https://doi.org/10.1038/s42003-020-01577-x

Nancy, J. L. (2020). *Communovirus*. Accessed on 11th May 2021. https://www.versobooks.com/blogs/4626-communovirus

Natural History Museum. (2020, September 21). Nature: Liberated by lock-down? *Natural History Museum*. Retrieved November 20, 2021, from https://www.nhm.ac.uk/discover/nature-liberated-by-lockdown.html

OECD. (2020). *Building back better: A sustainable, resilient recovery after COVID-19*. Available at: https://read.oecd-ilibrary.org/view/?ref=133_133639-s08q2ridhf&title=Building-back-better-_A-sustainable-resilient-recovery-after-Covid-19&_ga=2.104523456.1389674155.1632825099-113007773.1632825099

Pfizer. (2021). *Pfizer's novel covid-19 oral antiviral treatment candidate reduced risk of hospitalisation or death by 89% in interim analysis of phase 2/3 epic-HR study*. Pfizer. Retrieved November 20, 2021, from https://www.pfizer.com/news/press-release/press-release-detail/pfizers-novel-covid-19-oral-antiviral-treatment-candidate

Porat, T., Burnell, R., Calvo, R. A., Ford, E., Paudyal, P., Baxter, W. L., & Parush, A. (2021). "Vaccine passports" may backfire: Findings from a cross-sectional study in the UK and Israel on willingness to get vaccinated against COVID-19. *Vaccines, 9*(8), 902. https://doi.org/10.3390/vaccines9080902

Sabelli, C. (2021, March 25). Covid-19 apps are effective even with 20% uptake. *Nature News*. Retrieved November 20, 2021, from https://www.nature.com/articles/d43978-021-00034-5

Sawal, I., Ahmad, S., Tariq, W., Tahir, M. J., Essar, M. Y., & Ahmed, A. (2021). Unequal distribution of COVID-19 vaccine: A looming crisis. *Journal of Medical Virology, 93*, 5228–5230. https://doi.org/10.1002/jmv.27031

Schwab, K., & Malleret, T. (2020). *COVID-19: The great reset*. Amazon Digital Services LLC – KDP Print.

Sharon, T. (2021). Blind-sided by privacy? Digital contact tracing, the Apple/Google API and big tech's newfound role as global health policy makers. *Ethics and Information Technology, 23*(S1), 45–57. https://doi.org/10.1007/s10676-020-09547-x

Strauss, V. (2020). *The Washington post*. Accessed on 11th May 2021. https://www.washingtonpost.com/education/2020/05/06/cuomo-questions-why-school-buildings-still-exist-says-new-york-will-work-with-bill-gates-reimagine-education/

Surti, M., Patel, M., Adnan, M., Moin, A., Ashraf, S. A., Siddiqui, A. J., … Reddy, M. N. (2020). Ilimaquinone (marine sponge metabolite) as a novel inhibitor of SARS-CoV-2 key target proteins in comparison with suggested COVID-19 drugs: Designing, docking and molecular dynamics simulation study. *RSC Advances, 10*(62), 37707–37720. https://doi.org/10.1039/D0RA06379G

Szawarski, P., & Rich, C. (2021). Politicisation of ivermectin raises concerns about how we communicate with the public. *BMJ, 373*, n1258. https://doi.org/10.1136/bmj.n1258

Troncoso, C. (2021, February). Contact tracing apps: Engineering privacy in quicksand. Enigma, *USENIX*. https://www.usenix.org/conference/enigma2021/presentation/troncoso

Tupper, P., Otto, S. P., & Colijn, C. (2021). Fundamental limitations of contact tracing for COVID-19. *FACETS, 6*(1), 1993–2001. https://doi.org/10.1139/facets-2021-0016

Umbrello, S. (2021). Should we reset? A review of Klaus Schwab and Thierry Malleret's 'COVID-19: The great reset' [Book Review]. *Journal of Value Inquiry*, 1–8. https://doi.org/10.1007/s10790-021-09794-1

Venter, Z. S., Aunan, K., Chowdhury, S., & Lelieveld, J. (2020). COVID-19 lock-downs cause global air pollution declines. *Proceedings of the National Academy of Sciences, 117*(32), 18984–18990. https://doi.org/10.1073/pnas.2006853117

Open Access This chapter is licensed under the terms of the Creative Commons Attribution 4.0 International License (http://creativecommons.org/licenses/by/4.0/), which permits use, sharing, adaptation, distribution and reproduction in any medium or format, as long as you give appropriate credit to the original author(s) and the source, provide a link to the Creative Commons license and indicate if changes were made.

The images or other third party material in this chapter are included in the chapter's Creative Commons license, unless indicated otherwise in a credit line to the material. If material is not included in the chapter's Creative Commons license and your intended use is not permitted by statutory regulation or exceeds the permitted use, you will need to obtain permission directly from the copyright holder.

Part I
Learning from COVID-19

Part 1
Learning from COVID-19

Chapter 2
COVID-19 and Changing Values

Ibo van de Poel, Tristan de Wildt, and Dyami van Kooten Pássaro

2.1 Introduction

The coronavirus pandemic and the measures taken to mitigate its effects, such as lockdowns, have hugely affected people's lives. It seems likely, therefore, that it may have also affected people's values, at least in the short term. Our aim in this chapter is to explore whether the COVID-19 pandemic has led to value changes in society, and if so, how.

There have been a few studies addressing value change due to the COVID-19 crisis. Steinert (2020) addresses the possibility of value change due to what he calls emotional contagion through social media, which, according to him, may lead to more emphasis on values stressing security preservation and threat avoidance. Lampert et al. (2021) and Reeskens et al. (2021) report relevant results from value surveys. While the latter find that values remain largely stable, the former – among others – find that "[t]he pandemic and the economic crisis it brought have led to an increased focus on individual free choice and the non-material aspects of life. At the same time, the support for […] law and order have decreased. People are increasingly calling for inclusive growth and for reducing the gap between rich and the poor" (Lampert et al., 2021: 3). Similarly, Liscio et al. (2021) examine values in the COVID-19 pandemic, although they do not address value change.

The limited studies available also make seemingly contradictory speculative claims about how values (may) change due to the corona pandemic, from an increasing emphasis on security values (Steinert, 2020) to no value change (Reeskens et al., 2021) to more emphasis on post-materialist values (Lampert et al., 2021). Our study adds to this ongoing debate by studying possible value changes based on how news

I. van de Poel (✉) · T. de Wildt · D. van Kooten Pássaro
Delft University of Technology, Delft, The Netherlands
e-mail: I.R.vandePoel@tudelft.nl; T.E.deWildt-1@tudelft.nl;
D.vanKootenPassaro@student.tudelft.nl

© The Author(s) 2022
M. J. Dennis et al. (eds.), *Values for a Post-Pandemic Future*, Philosophy
of Engineering and Technology 40, https://doi.org/10.1007/978-3-031-08424-9_2

reports on the corona pandemic. We analysed a large number of such articles about the COVID-19 crisis from six countries (US, UK, India, South Africa, Japan, South Korea) to trace how often certain values were addressed. Additionally, we looked at news articles from 2016 to early 2020 to see how the COVID pandemic might, or might not, have influenced the frequency with which certain values are addressed in new articles compared to the pre-COVID period. Our analysis looked at eleven different values: health and safety, economic welfare, mental health, socio-economic equality, freedom, democracy, sustainability, privacy, conformity, family and belonging, and hedonism.

To analyse this large set of news articles, we employed a computational tool: topic modelling, which allows tracing the changing frequency of specific topics in a text corpus. For several methodological reasons, topic modelling is likely to provide a more reliable analysis of values, and value changes, than a keyword-based counterpart (de Wildt et al., 2021). However, as we will explain, care should be taken in interpreting the results of such analyses, as what we find are changes in the frequency of references to certain values, which leaves open the question of what such changes signify and whether they truly reflect the importance people attach to values in their lived lives. Moreover, we remain open to the possibility that the way in which we construed the value topics in our computational topic model may not always fully or adequately reflect the values we are interested in.

We proceed as follows. Section 2.2 gives some background on the notion of 'value' and introduces the eleven values we have analysed. Section 2.3 explains our methodology. Section 2.4 presents the main results. Section 2.5 discusses possible interpretations of these results. We finish by elaborating on these various interpretations in our conclusion.

2.2 Values in the COVID-19 Pandemic

2.2.1 What Are Values?

Values are generally taken to be expressions of what is 'good' or 'desirable'. However, beyond this general consensus, there are marked differences in how different disciplines and scholars have understood the term 'value' and how they have understood value change. Therefore, before discussing relevant values – and possible value changes – for the COVID-19 crisis, we will start with a brief overview of the notion of value as it has been roughly understood in psychology, sociology and (moral) philosophy.

Psychologists usually view values as part of an individual's personality (Steg & De Groot, 2012). They are often taken to be beliefs about what is, in general terms, desirable (Rokeach, 1973; Schwartz, 1992). Furthermore, values are seen as abstract, general, and relatively stable over a person's life. Like Schwartz (1992), some psychologists take values to be universal, although their relative importance may change over time and between nations and cultures. Examples of values distinguished by Schwartz are benevolence, achievement, and security.

In addition to this more psychological notion of value, one might distinguish a more sociological one, which understands value as a social phenomenon or cultural resource (cf. Demski et al., 2015; Jasanoff & Kim, 2015). Understood in this way, values are shared anchors that people use to justify their behaviour to others and to which they orient their actions to a greater or lesser extent. For example, generosity may be a social value in the sense that in a specific society or community, people expect each other to be generous to one another. Such social values may be influential even if they deviate from the more personal values distinguished by psychologists. This is because they typically express social expectations about how others will behave and what behaviour others will – and will not – accept. So even people that do not have a generous personality may behave generously because others expect them to do so. Although such social values may be stable over long time periods, they may also change; new values may emerge, etc. Moreover, there is usually some room for agents to (re)interpret these values and their meaning and what they imply for the desirability of certain actions or technologies.

A third relevant notion of values is that of moral values. Moral values express what is normatively or morally good and desirable. For example, fairness is often considered a moral value. In (moral) philosophy, there are many different (metaethical) accounts of values. Still, an important distinction is between accounts that associate values with (subjective) mental states like desires and accounts that take values to be objective and real in some sense. However, even most accounts that associate values with desires do not equate them with actual desires. Instead they associate them with, for example, informed desires or desires under certain conditions. Concretely this can be something like seeing the world from behind a veil of ignorance about one's specific position in society (e.g., Rawls, 1999 [1971]).

In the remainder of this chapter, we will use the terms 'personal values', 'social values' and 'moral values' to refer to these three different types of value. It should be stressed that our usage of these terms connate different uses of the term 'value', not necessarily distinctions in the content of a value. Thus, sustainability can be a personal as well as a social or a moral value. Moreover, it can be all three simultaneously. This is because three usages of the term 'value' are not necessarily conflicting, but rather refer to different phenomena; namely, a person's personality ('personal value'), shared anchors in society ('social value'), and expressions of what is morally good and desirable ('moral value').

2.2.2 Relevant Values for the COVID-19 Pandemic

For the methodology we have used in this chapter (explained in Sect. 2.3), we used a computational tool to trace values and value changes in large text corpora, in this case, news articles about COVID-19. This approach is particularly appropriate for tracing social values, as it seems likely that news articles would refer to shared values in society to a more significant extent than, say, their authors' personal values or moral values. Still, it would seem reasonable to assume that the values we find this way also tell us something about a population's personal values and what members of this population consider to be morally important. The latter is not necessarily the same as moral values, of course, but is often a proxy for them.

In making an inventory of relevant values, we have first brainstormed together (as authors) on what the relevant (social) values in the COVID-19 crisis could be. Additionally, we have used the results of a study by Liscio et al. (2021), who let two teams of human annotators identify values in text corpora based on a PVE (Participatory Value Evaluation) study on relaxing COVID-19 measures in the Netherlands (Mouter et al., 2021). This resulted in the addition of three values; see Appendix 2 for details. Below, we briefly give a short explanation of each value and justify why we consider these values relevant. We do not claim that our list of relevant values is exhaustive, although we believe it is relatively comprehensive.

Health and safety: Health has been defined by the World Health Organisation (WHO) as the "state of complete physical, mental and social wellbeing and not merely the absence of disease or infirmity" (World Health Organization, 2006: 1). Safety may be understood as the absence – or at least the reduction in as far as reasonably possible – of risks, in this case mainly health risks. Health and safety are obviously relevant: at the moment of writing, there are almost 4 million confirmed deaths worldwide due to COVID-19, with actual numbers likely reaching much higher due to limited testing and attribution difficulties (Ritchie et al., 2020).

Economic welfare may be understood as the level of prosperity and the standard of living of a country or individual. We understand it here primarily in economic terms, and the value is therefore different from a value like wellbeing. The pandemic is estimated to lead to a loss in global GDP (gross domestic product) of 4.5% in 2021, equaling around 3.94 trillion US dollars in lost economic output (Szmigiera, 2021). As soon as May 2020, 30% till 35% of respondents in Germany, the UK and the US reported a loss in income due to corona[1]; Eurostat reports a loss in median income in the EU in 2020 of 5.2% compared to 2019.[2]

The WHO defines *mental health* as "a state of well-being in which every individual realises his or her potential, can cope with the normal stresses of life, can work productively and fruitfully, and is able to make a contribution to her or his community."[3] While it might be argued that the value of 'mental health' is part of the value 'health and safety', we distinguish it here as a separate value because it denotes quite specific considerations. Some of the measures deemed necessary to achieve health and safety, like lockdowns, are detrimental to mental health. In a US health tracking poll in July 2020, 53% of the respondents reported a negative impact

---

[1] https://www.statista.com/statistics/1108061/losing-income-due-to-the-covid-19-corona-pandemic/. Accessed 22-5-2021.

[2] https://ec.europa.eu/eurostat/web/products-eurostat-news/-/ddn-20201210-2.   Accessed 22-5-2021.

[3] https://www.who.int/news-room/fact-sheets/detail/mental-health-strengthening-our-response. Accessed 22-6-2022.

on their mental health.[4] Similarly, the share of adults in the US reporting symptoms of anxiety disorder and/or depressive disorder raised four-fold (from 11% to 41%) between January–June 2019 and January 2021.[5]

In this paper, *socio-economic equality* is understood as equality between different social groups, including differences in race, gender, age, and between nations. It relates to equality of opportunity but also equality of outcome (e.g., income). There are numerous signals that both the impact of COVID-19, as well as those of countermeasures, is unequally distributed over the population in many countries, as well as worldwide. In many cases, the vulnerable and already disadvantaged groups take on the most significant part of the burden (Perry et al., 2021; Clouston et al., 2021; Cifuentes et al., 2021; Lopez et al., 2021).

Freedom may be understood as the ability to direct one's life (autonomy), but it is also often understood as the absence of external constraints and hindrances. The latter seems particularly relevant in the COVID-19 crisis, which has a considerable impact on freedom due to social distancing, lockdowns, night curfews and bans on (large) gatherings.

Democracy as a value does not only refer to a particular mode of government, but also to equal access to a number of democratic and human rights as well as respect for the rule of law and political equality. Unfortunately, democratic values have come under pressure because slowing the spread of the SARS-CoV-2 virus has required extraordinary governmental measures that can be hard to publicly justify in a democracy. According to a report from the Freedom House, the conditions of democracy and human rights have worsened in 80 (out of 192) countries during the pandemic (Repucci & Slipowitz, 2020).[6]

Environmental sustainability refers to the value of sustaining environmental resources and reducing environmental pollution and degradation. For example, the pandemic is reported to have led to a reduction in greenhouse gas emissions and improved (local) air and water quality, as well as to an increase in medical waste and consequent shoreline pollution (Cheval et al., 2020; Bhat et al., 2021; Rume & Didar-Ul Islam, 2020; Rupani et al., 2020).

Privacy in this context is understood as the protection of the personal sphere against intrusion by others. For the COVID-19 pandemic, informational privacy, which refers to the ability to decide what information about oneself to share with others or keep confidential, is essential. Privacy is particularly an issue because of the privacy risks of COVID-19 tracing apps and home monitoring technology (Chan & Saqib, 2021; Gerke et al., 2020).

Conformity is understood here in terms of the population's willingness to abide by anti- COVID-19 measures, mainly from governments. Schwartz situates

---

[4] https://www.kff.org/coronavirus-covid-19/report/kff-health-tracking-poll-july-2020/. Accessed 22-5-2021.

[5] https://www.kff.org/coronavirus-covid-19/issue-brief/the-implications-of-covid-19-for-mental-health-and-substance-use/. Accessed 22-5-2021.

[6] The cited study is based on a survey among 398 experts from 105 countries and additional field and desk research.

obedience as being motivationally close to values like conformity and tradition, both of which relate to the subordination of the self to social expectations (Schwartz, 1992). It has, however, also been suggested that obedience to COVID-19 rules is not only (or primarily) to be explained in terms of conformity and authority, as it might also be based on a perception of procedural justice (Reicher & Stott, 2020).

Family and belonging is the value of being part of – and deriving part of one's identity from – a larger social group, like one's family, friends, neighbourhood, cultural group, or nation. Because of anti-COVID-19 measures, some important social ties for belonging like work, school, or the university, have been weakened. Meanwhile, others, in particular the family, may have been strengthened.

Hedonism. In moral philosophy, hedonism is the theory that equates the value of human wellbeing with pleasurable experience. Similarly, psychologists associate hedonism with excitement, pleasure, new experiences, and self-indulgence (Schwartz, 1992). However, COVID-19 has obviously made such activities more difficult. During the pandemic, many have found it difficult to express their hedonistic values, which may have resulted in more emphasis on other values and/or a negative impact on mental health.

2.3   Method

2.3.1   Topic Modelling as a Method to Trace Value Change

Values tend to be discussed in a latent manner in text corpora. Rather than explicitly naming the value in question, authors often use a wide range of words for referring to a value. For example, when an author discusses the impact of COVID-19 on the current energy transition, the probability that the author explicitly mentions the value 'environmental sustainability' is limited; the author might use such words as 'renewable', 'durability' and 'planet' to refer to the idea of environmental sustainability. The fact that values tend to be discussed in a latent manner has implications for how value change can be studied in text corpora. Studying value change using topic modelling typically requires a large number of texts to ensure that the trends observed are not arbitrary. Furthermore, using many texts calls for the use of keywords instead of a manual analysis to identify those texts which are addressing values of interest.

Nevertheless, the fact that values are latent means that it is difficult to find a set of keywords that matches the idea of a value (de Wildt et al., 2021). The set of words used by authors to refer to a value can be considerable. Some of these words (e.g. 'durability' and 'planet') may not be related to environmental sustainability when used in different contexts (e.g. material sciences or planetary science). Using only the relevant value term (like 'environmental sustainability') as a keyword typically leads to underestimating the number of texts addressing this value, while adding more keywords might lead to overestimating it.

A limited number of approaches exists in the academic literature to study values in text corpora. Liscio et al. (2021) propose the 'Axes' approach, which helps identify context-specific values and related keywords. While complemented by Natural Language Processing, the process still relies on human annotation to identify values, which may be time-intensive. Sun et al. (2014) propose an approach entitled Automatic Estimation of Schwartz Values (AESV). This approach focuses on identifying Schwartz values (Schwartz, 1992) in social media and can calculate the value proprieties of individuals and groups. Similarly, de Wildt et al. (2018) propose an approach based on probabilistic topic models (Blei & Lafferty, 2009) to capture the gist of text corpora that address values. This approach, further refined by de Wildt et al. (2021), is used here.

Using probabilistic topic models, values are defined using distributions of words instead of keywords. Probabilistic topic models originate from the field of text mining. In a topic model, a topic is defined as a distribution of words. For example, a topic on vaccines as measures against Covid-19 might have high probabilities on terms such as 'RNA' and 'shot' and low probabilities on 'mask' and 'hand'. The construction of a topic model can be done in an unsupervised or semi-supervised manner: In the first case, resulting topics will tend to converge to the most frequent themes in the text corpus. In the second case, topics can be shaped so that they represent some themes of interest, like – in our case – values. Texts addressing values can then be identified by comparing the distribution of words in a text and the distribution of words of topics built to reflect the idea of specific values.

A number of potential biases need to be considered when using probabilistic topic models to trace value change (cf. de Wildt et al., 2021). On one side, probabilistic topic models allow for better capturing the idea of a value in comparison to keywords. The dataset analyzed can be large, thereby helping to explore a wider set of sources expressing different perspectives. On the other side, the type of corpus analyzed might affect the type of values identified and the way they are discussed. For example, newspaper articles often focus on human values while a corpus composed of patents might concentrate on technical ones. Also, the time length of the dataset might affect the type of value change observed (e.g. temporary punctuated shock or durable value change). Finally, we use the frequency of occurrence of values in texts as a proxy for the (relative) importance of values. We discuss how to interpret topic model outcomes given these potential biases in Sect. 2.3.2 ('Interpreting outcomes'). We reflect further on these biases when interpreting topic model results in Sect. 2.5.

2.3.2   Data Collection and Analysis

The process of exploring value change using topic models involves three steps: (1) selecting the dataset, (2) choosing the number of topics to search and (3) creating topics that represent the relevant values (de Wildt et al., 2021). The topic model created can also be exported and applied to new datasets. This section describes how each step has been used for this research. We also discuss how to interpret model outcomes, i.e. what frequencies mean regarding the importance of values. The datasets and notebook used for this analysis can be found online.[7]

Selecting the Dataset

To pull from a robust set of articles for the topic modelling analysis, we have used the following four guidelines for finding and using datasets for this research:

First, we looked for text sources that could help trace potential value change occurring from the start of the COVID-19 crisis until the time of analysis. We have selected newspaper articles as they are expected to depict important values in society. In contrast, the typically long publication process of scientific articles might not allow observing value change occurring within a timeframe of several months. However, different types of text corpora might concentrate on different values, and their analysis might depict different kinds of value change (de Wildt et al., 2021). We have considered this in the interpretation of our results and discuss this limitation in Sect. 2.6.

Second, we looked for both datasets that are specifically on COVID-19, allowing us to explore value change in dealing with the COVID-19 crisis, as well as datasets not explicitly related to COVID-19, allowing us to explore how the crisis has affected overall values in society.

Third, the datasets need to be sufficiently large to ensure representativeness. The minimum number of texts depends on the length of the timeline analysed and the precision of the analysis required. For most analyses, a minimum number of 1000 texts is required.

Fourth, the sources needed to be in English as a topic model would typically not be able to form one topic if it is discussed in different languages due to semantic differences.

The following three datasets of news articles were ultimately used for this research. A detailed overview of the datasets, including the number of news articles and newspaper sources, is provided in Appendix 1.

- A corpus with news articles on the COVID-19 pandemic from the United States (US), United Kingdom (UK) and South Africa for the period January 2020 – August 2020. This corpus is drawn from a dataset from Aylien Ltd. (2020), from which we have extracted 5000 randomly selected news articles for every country mentioned.
- A corpus with news articles on the COVID-19 pandemic from January 2020 – January 2021 from Japan, India and South Korea. This corpus is based on a dataset collected by Ghasiya and Okamura (2021).
- A corpus with text articles from Reuters (category 'world news') for the period January 2016 – March 2020. This corpus also contained news articles not related to COVID-19. This corpus is based on a dataset from Thompson (2020), from which we selected all articles with category 'world news.'

---

[7] https://doi.org/10.4121/20134163

Choosing the Number of Topics to Search

The creation of a topic model requires indicating the number of topics that the algorithm needs to find. The number of topics should be sufficiently large to ensure that enough space is given to semi-supervised topics created (e.g. representing values) and other topics occurring in the dataset to converge to. However, an excessively high number of topics (e.g. 1000 topics) will vastly increase the time required by the algorithm to create the topic model. Therefore, we have set the number of topics to 200 and have verified that this number was sufficient to develop topics that represent relevant values for the COVID-19 pandemic.

Creating Topics that Represent Values

Creating topics that represent values is a process of pushing and pulling anchor words to ensure that each distribution of words formed adequately represents the relevant value (de Wildt et al., 2021). Anchor words are words used as input to a semi-supervised topic model and help steer the topic in a particular direction (i.e. a specific distribution of words). For example, the words 'health', 'safety', 'death' and 'immune' can be used to create a topic for the value *health & safety*. However, in case the newly created topic still includes aspects that are not related to *health & safety*, these unrelated words can be used as anchor words to create a separate topic, hereby pulling out this aspect from the topic on *health & safety* into a separate topic and specifying the topic of *health & safety* to suit our understanding of the value.

Table 2.1 provides an overview of the anchor words used to create topics representing values. The column 'Topic created' shows the 10 highest probability words for the distributions of words formed for each topic.

We have verified the quality of topics in two ways: we have manually verified that documents assigned to topics on values were indeed addressing the values in question as well as verified that none of the topics not related to values still contained aspects of values by looking through the list of all generated topics.

Interpreting Outcomes

In interpreting the outcome of the analysis, three important considerations should be borne in mind:

First, the analysis performed reports about frequencies (i.e. the percentage of newspaper articles addressing a value at a specific moment in time). At the same time, we are ultimately interested in changes in the importance of values. For example, the fact a value is named more frequently might be caused by an emerging problem concerning this value (for example, a new technology that creates a moral issue), as well as by a technical or regulatory solution that has been found to better address this value (e.g. a new COVID-19 vaccine). Thus, while changing frequencies of values might be signs of changes in importance, a further reflection about what could have caused changes in frequencies is essential before conclusions can be drawn.

Second, uncertainty always exists about the quality of topics. This is particularly the case for the topic of values, since values are sometimes hard to separate semantically from how they are being operationalised (e.g. the system used to act upon them). An example of this is 'democracy', which strongly refers to both a value and

Table 2.1 Anchor words and topics created (10 most prominent terms displayed) that represent values

Values	Anchor words	Topic created
Health & Safety	Safety, health, healthy, deaths	[Health, deaths, public health, the health, of health, health and, public, health minister, safety, health officials]
Mental Health	Isolation, depression, suicide, solitude, somber, anxiety, sadness, mental health	[Isolation, anxiety, mental health, depression, self-isolation, in isolation, self, mental, sadness, suicide]
Economic Welfare	Economic, costs, cost effective, stimulus, bankruptcy, debt	[Economic, stimulus, debt, economy, billion, costs, financial, market, business, the economy]
Socio-Economic Equality	Equality, equal, fairness, socio-economic, socio-economic class, inequality, unequal, working class, equity, income differences, living standard, insecurity, divide	[Equal, equity, inequality, equality, divide, insecurity, working class, unequal, toward, policies]
Privacy	Privacy, private, personal, secret, tracking, invisible, security, monitoring	[Personal, private, security, monitoring, tracking, privacy, and private, of personal, security and, and personal]
Freedom	Freedom, choice, autonomy, personal responsibility, independence	[Choice, freedom, freedom of, power, speech, independence, reality, diverse, views, no choice]
Democracy	Choice, suppression, public opinion, opinion, rights, totalitarian, authority, democracy	[Rights, opinion, democracy, political, leaders, human rights, legal, authority, society, human]
Environmental Sustainability	Sustainability, sustainable, renewable, durability, climate change, global warming, pollution, environment, environmental, air pollution, water quality	[Environment, sustainable, environmental, climate change, sustainability, pollution, climate, creation, the environment, the creation]
Hedonism	Enjoyment, pleasure, wellbeing, friendship, pleasurable, enjoy, stress, self-esteem, fun, hobby, new experience, experience, sports, pub, alcohol, conviviality, entertainment, enjoy, positivity, outdoors, leisure, joy, partying	[Experience, entertainment, stress, fun, enjoy, joy, outdoors, pleasure, friendship, positivity]
Community and Family	Community, family, belonging, group, relatives, friends, friend, children, neighbour, neighbours, neighbor, neighbors	[Family, children, friends, relatives, friend, parents, the family, family and, his family, friends and]
Conformity	Conformity, restriction, follow the rules, obedience, conventional, law, order, obedience, norms, culture, heritage	[Order, law, order to, in order, culture, the law, law enforcement, enforcement, restriction, home order]

a system of government. This potential bias does not prevent comparisons of frequencies of one value between multiple countries, as this bias is likely to be the
same for every country. Neither does it hamper a qualitative comparison of patterns
of value frequencies within and between countries, as this bias is the same over the
timeline of the dataset. However, a numerical comparison between values – for
example stating that one has become more frequent than the other – should be
treated with care. The validity of such a comparison would depend on the extent to
which both topics genuinely represent the value they aim to represent.

Third, the choice of the datasets was primarily based on availability. As it was
very hard to find (publicly available) relevant datasets, we decided to reuse datasets
collected by others (see Appendix 1). This means that we could not ourselves ensure
the representativeness of the datasets nor correct for potential biases in the dataset
(e.g. partisan views in the US). Nevertheless, we have no reason to assume that the
datasets are not representative or biased; but obviously caution should be taken in
the interpretation of the results for this reason.

2.4 Results

Here we present the main results of our analysis of how the frequency of specific
values has changed over time in different countries compared to the pre-corona
period. Figures 2.1, 2.2, 2.3, 2.4, 2.5, and 2.6 show the results of the six countries
we have analysed. Please note that the time span is somewhat different for the different countries due to the (dis)availability of data.

Based on the results for these six countries, we make the following observations:

1. At the start of the pandemic (January–February 2020), the value of *safety and
   health* is addressed in at least 60% of the news articles in all six countries, with
   somewhat higher frequencies for Japan and India (up to 80%). However, in all
   countries, this percentage drops to about 40% from April–June 2020 and then
   stabilises.
2. The general pattern for the other values seems to be that the trend in frequency
   goes up over time for most of them. However, there are distinct differences
   between countries and values here (see our successive observations). What is
   also worth observing is that in South Korea, as early as April 2020, at least one
   other value becomes as frequent as *safety and health*, while in India this takes
   until the end of 2020.
3. Concerning the value of *economic welfare*, we see three different patterns:

   1. In Japan and South Korea, we see a considerable increase in frequency until
      April 2020 and then a stabilisation at a relatively high level (around 35–40%).
   2. In both the US and the UK, we see a peak in frequency in March 2020 (around
      30%) and then a stabilisation at a lower level (approximately 25% in the US
      and 20% in the UK).

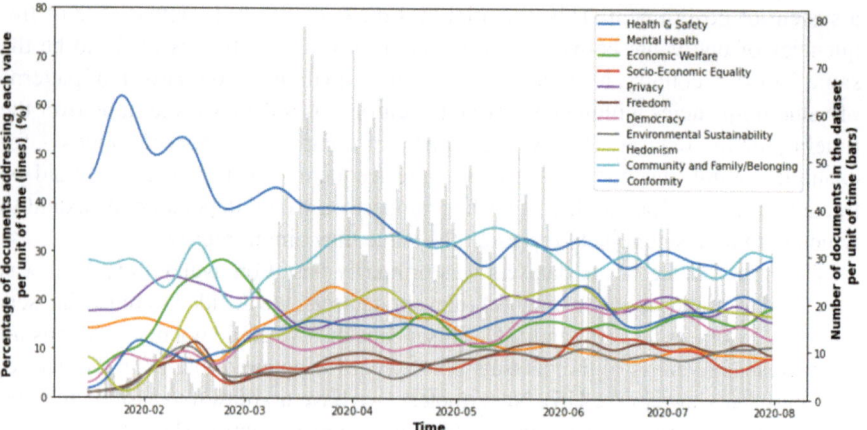

Fig. 2.1 United Kingdom

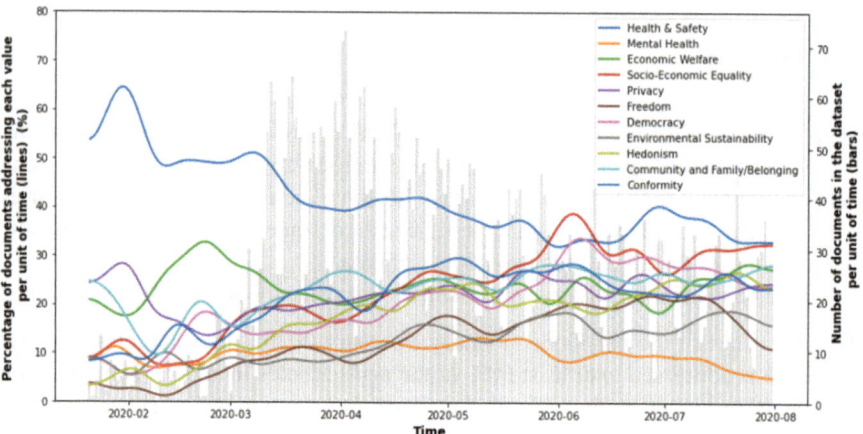

Fig. 2.2 United States

3. In South Africa and India, we see an increase over time at a relatively low overall level of frequency (around 10–20%).

4. While we see an increase in frequency over time for the values of *democracy* and *privacy* in all six countries, the growth is most marked in South Africa (up to around 40% in August 2020) and India (approximately 30% in the second half of 2020).

5. In the US, we also see a marked increase in the value of *socio-economic equality* frequency from about 10% in early 2020 to around 40% between June and August 2020. In other countries, we also witness an increase in the frequency of this value over time, but at a slower pace and never reaching quite such a high percentage.

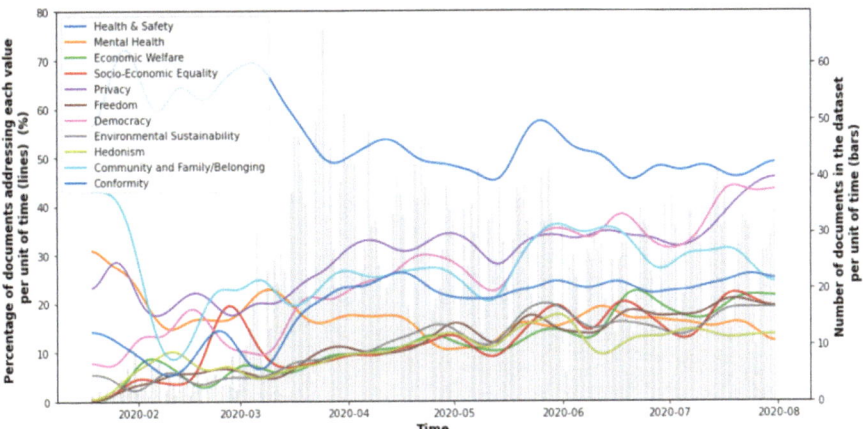

Fig. 2.3 South Africa

6. Although the value of *hedonism* seems to increase in frequency in all six countries, it goes up most markedly in the US, where it rose from below 10% in January 2020 to around 30% in August 2020. In contrast, it tends to go up less steeply in the other countries, rising from approximately 10% to only about 20%.
7. Concerning the value *conformity*, we observe that the frequency increases in the US, the UK and South Africa while remaining relatively stable in other countries.

Figure 2.7 shows the results for the corpus with new articles in the period 2016–20, including non-COVID news. It very clearly shows the effect of the COVID-19 pandemic on the frequency in which specific values are addressed in news articles. *Health & safety* increase from below 10% to above 50% in three months. *Hedonism*, *mental health* and *economic welfare* also show an increase in frequency in early 2020, although the frequency of these values does not deviate from their bandwidth in the period before 2020.[8] The other values show a drop in frequency. For *democracy*, *privacy* and *socio-economic equality*, this is a drop well below the bandwidth of the values in the period 2016–20.

2.5   Discussion

We discuss the following four points:

• The general pattern of value change and whether we can expect any long-term value changes due to the COVID-19 pandemic.
• Possible explanations for differences between countries we found.

---

[8] Given the large number of articles in this dataset, this bandwidth would seem a reliable indication for 'normal' variations in the frequency of values in the pre-COVID time span.

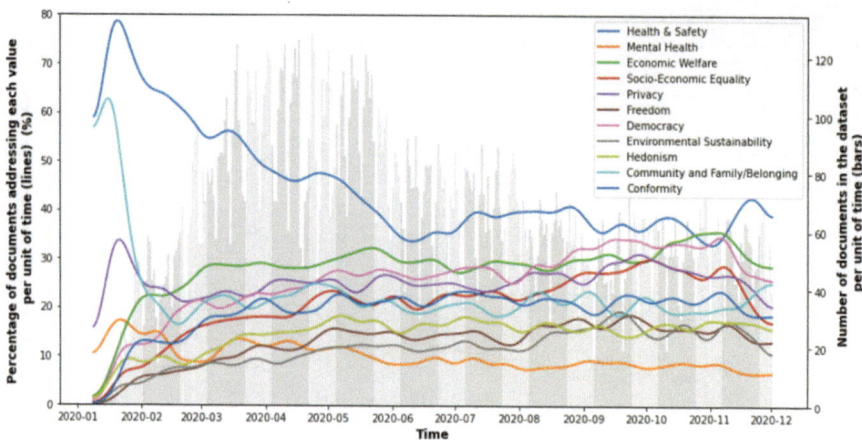

Fig. 2.4 Japan

- A comparison of our results with what might be expected based on existing value theories.
- Potential moral implications of our findings.

General Pattern of Value Change

As Fig. 2.7 shows, the COVID-19 pandemic has led to a punctuated shock in the frequency in which certain values are addressed in news articles. In particular, the value of health and safety went up rapidly in frequency at the start of the pandemic. We also see that some other values at stake in – or somewhat threatened by – the pandemic, such as mental health, economic welfare, and hedonism, go up in frequency, although this increase is not significant compared to previous fluctuations in the 2016–20 period. Conversely, the frequency of all other values drops, particularly for democracy, privacy, and socio-economic equality. An explanation for this may be that these values are not, or at least not immediately or initially, associated with COVID-19 .[9]

When it comes to the long-term effect we might expect from this punctuated value change, the earlier observations 1 and 2 are significant. Together, they suggest that the impact of the punctuated value change we see in Fig. 2.7 in the first three months of 2020 is already *cancelling out* in the following months of the pandemic.[10] Thus, although it is hard to say anything definitive about whether the pandemic will lead to long-term value change, the pattern we can already witness *during* the

---

[9] One thing that should also be kept in mind is that if one value goes drastically up in frequency, like in this case health and safety, other are likely to go down as the amount of news articles will typically remain rather stable and articles will often address a limited number of values.

[10] Here, it should be kept in mind that the country trends in Figs. 2.1, 2.2, 2.3, 2.4, 2.5, and 2.6 are based on COVID news articles, not on all news articles, so that percentages cannot be directly compared with those in Fig. 2.7.

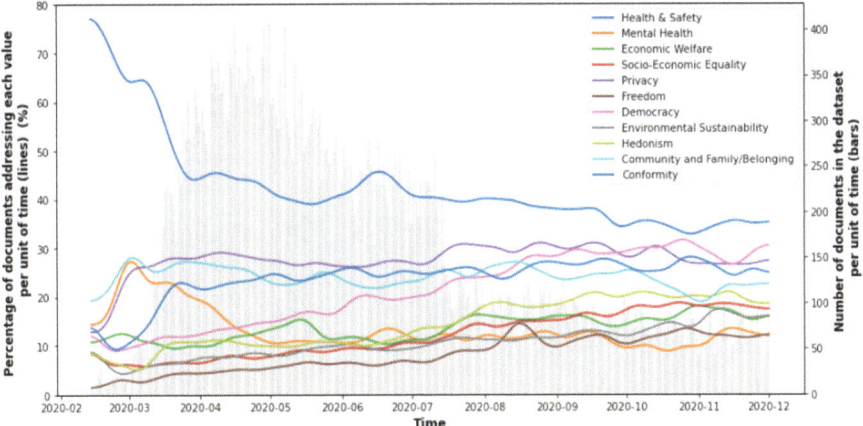

Fig. 2.5 India

pandemic suggests that the long-term effects on values may well be limited.[11] Instead, the pandemic may have led to punctuated shock reflected in a temporary change in the frequency in which certain values are addressed in news articles, which may smoothen out over time. Only time will tell whether this is really the case or whether there are also more enduring long-term effects.

Possible Explanations for Differences Between Countries

Here we look for possible explanations for observations 3–7. To do so, we referenced the following additional data for these countries to find possible explanations:

- COVID-19 cases and deaths (see Appendix 4).
- Stringency of measures (see Appendix 5).
- GDP per capita (Fig. 2.8), decline in GDP during corona (Fig. 2.10) and GINI coefficient (Fig. 2.11).
- Hofstede cultural dimensions (Fig. 2.9).

The first two of these additional data do not seem to correlate (in interesting ways) with the frequency of values in news articles; GDP data and the Hofstede dimensions seem relevant in some respects, as we will explain below.

Concerning the value of *economic welfare*, we observed three different trends in three groups of countries, i.e. (1) South Africa and India, (2) Japan and South Korea, and (3) the UK and the US (see observation 3 above). It is noteworthy that these three groups of countries have certain commonalities and, therefore, possibly each represent a larger group of countries. For example, (1) South Africa and India are

---

[11] Of course to say so, we would need to look at a dataset that also includes non-COVID news. Regretfully we have such a dataset only for the period until early 2021. Nevertheless, the trend we witness in the dataset with only COVID news suggests that the initial change in values may well be cancelled out over time, but to say anything more definitive we would need to know how this affects all news, not just COVID news.

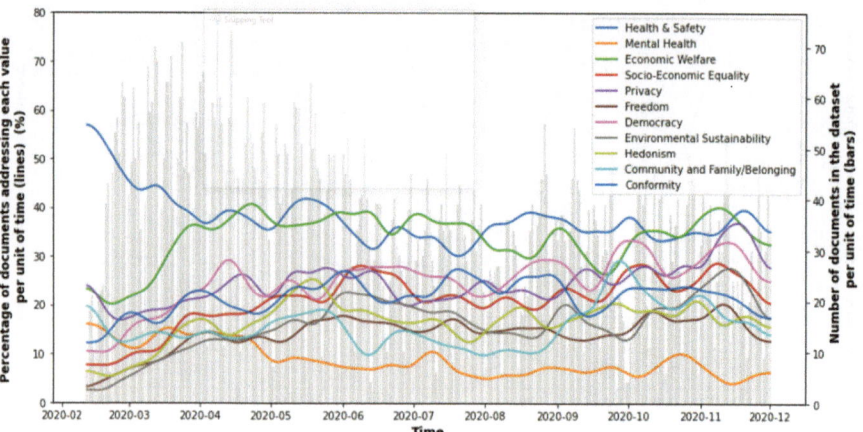

Fig. 2.6 South Korea

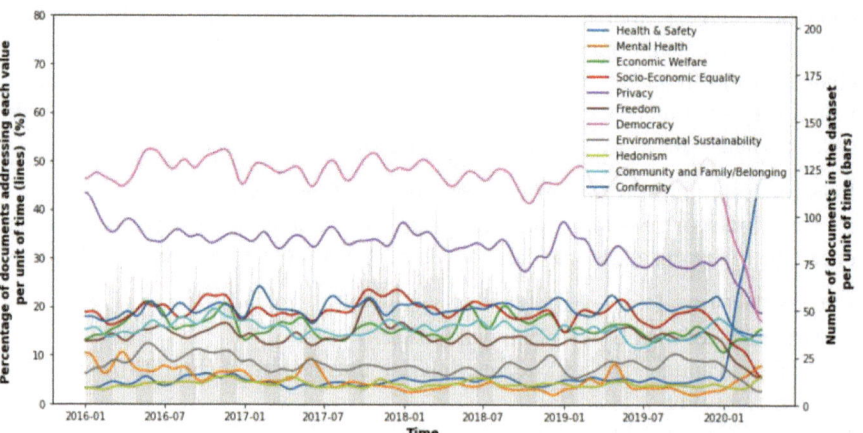

Fig. 2.7 Reuters World new also including non-COVID news

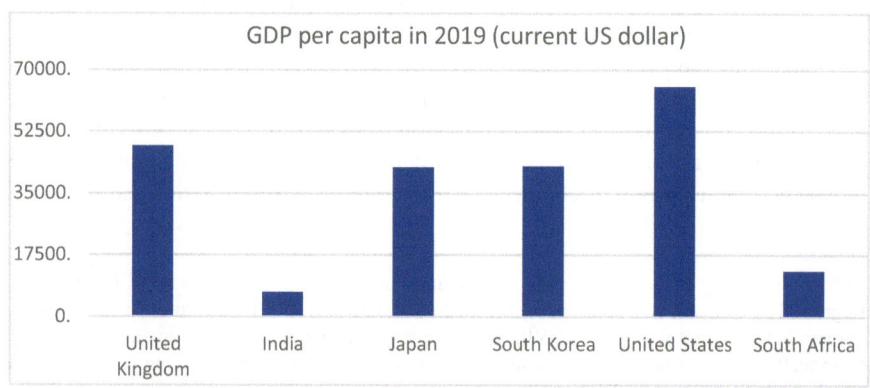

Fig. 2.8 GDP per capita in 2019 in current US Dollar. Data are from the World bank. Retrieved from https://data.worldbank.org/indicator/NY.GDP.PCAP.PP.CD at 1 July 2021

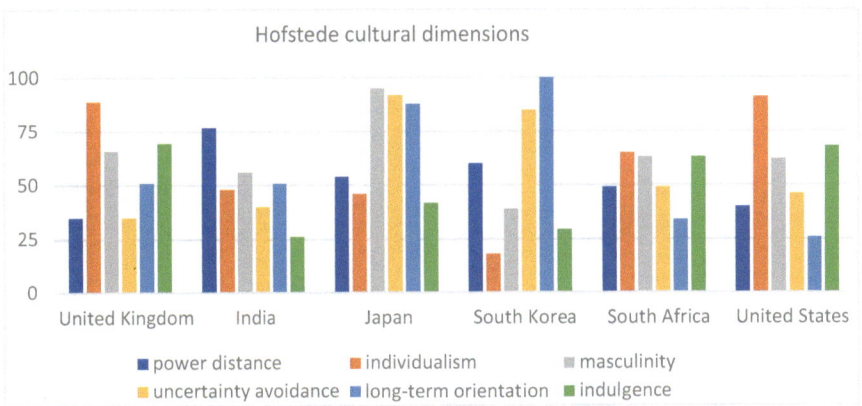

Fig. 2.9 Hofstede cultural dimensions. Data are based on Hofstede et al. (2010). Retrieved from https://geerthofstede.com/research-and-vsm/dimension-data-matrix/ at 1 July 2021

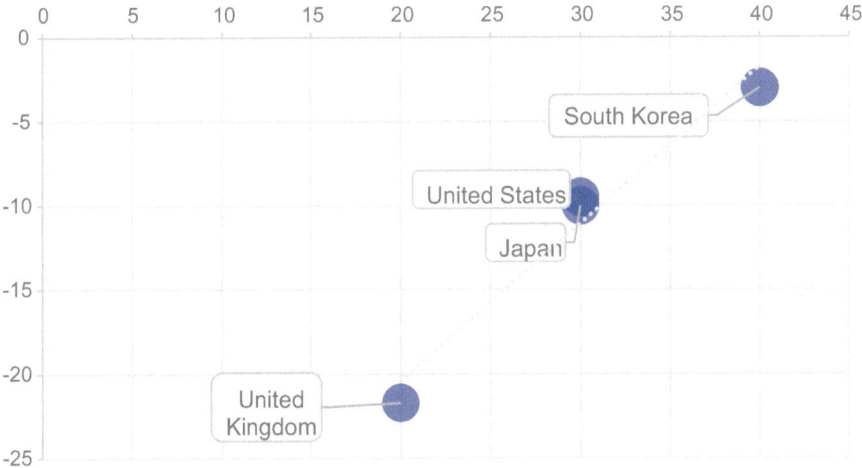

Fig. 2.10 Correlation between frequency of value 'economic welfare' and the percentual loss of GDP in four countries. The horizontal axis shows the frequency of the value 'economic welfare' in August 2020 in our data, and the vertical axis the percentual decrease in GDP in Q2 2020 (compared to Q2 2019). No data were available about decrease in GDP for South Africa and India. The GDP data were retrieved from https://ourworldindata.org/covid-health-economy on 1 July 2021

both countries from what has been called the Global South characterised by relatively low GDP per capita, (2) Japan and South Korea are both high-income countries from Asia, culturally characterised by a high uncertainty avoidance and a high long-term orientation and (3) the US and UK are both Western high-income countries, culturally characterised by a low uncertainty avoidance and a low long-term orientation.

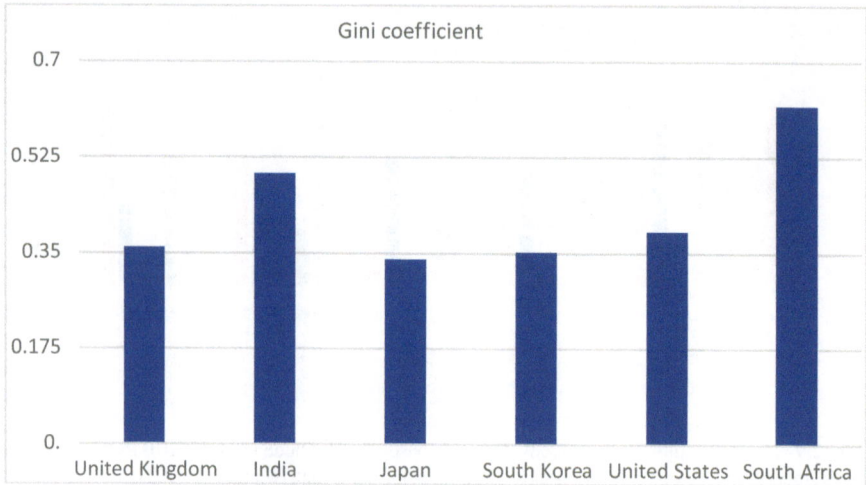

Fig. 2.11 Gini coefficient. 0 means total equality and 1 total inequality. The data are from 2015, except for India, which are from 2011. Data are from the OECD website. Retrieved from https://data.oecd.org/inequality/income-inequality.htm on 8 July 2021

Asian high-income countries, like South Korea and Japan, are likely most active in pursuing economic policies to abate the negative economic consequences of the pandemic; this may be (partly) explained by the cultural dimensions of (high) uncertainty avoidance and (high) long-term planning. Like the UK and US, Western high-income countries may also pursue such economic policies, but due to lower long-term planning and lower uncertainty avoidance, they may well less actively pursue such policies. Countries from the Global South, like India and South Africa, may lack the material means to afford such economic policies.

This suggests that the frequency of the value 'economic welfare' does not reflect how hard a country is hit economically by the pandemic, but rather how active it is in abating its adverse economic effects. This possible explanation is supported by Fig. 2.10, which shows a negative correlation between how hard certain countries are hit economically and the frequency of the value of 'economic welfare' in new articles.

When it comes to the values of *privacy and democracy* (observation 4), it is remarkable that the frequency of these values dramatically rises in countries from the Global South (South Africa and India). We do not have an explanation for this, but it belies the idea, sometimes heard[12], that such values may be considered less

---

[12] For example, Inglehart (2018) suggests that postmaterialist values (like privacy and democracy) are less prominent under conditions of scarcity. See also Inglehart and Welzel (2009). Also others scholars have suggested a correlation between economic development and democracy, although there is no agreement on the strength of the relation and in what direction it works (see e.g., Kauffman, 2021).

relevant in countries with low-income levels; we observe the opposite correlation.[13]

Concerning the value of *socio-economic equality* (observation 5), the US seems to be the exception because the frequency of this value increases here much more steeply than it does in the other countries. We do not see a clear correlation of the value with the GINI coefficients – a measure for income inequality – of the various countries, although the US does have a slightly higher GINI coefficient – i.e. more inequality – than the other high-income countries (Fig. 2.11). A better explanation for the trend in the US is perhaps the BLM (Black Lives Matter) movement which gained traction after the killing of George Floyd on 25 May 2020, immediately before we saw a peak in this value in the US in June 2020. This could be an effect from the news articles about these events being included in our sample (which is possible if they also contain COVID-19 keywords).[14] Furthermore, BLM may have increased awareness of racial and socio-economic inequalities, indirectly influencing the frequency of socio-economic equality in the dataset.

Concerning *hedonism* (observation 6), cultural differences may partly explain why we see a greater increase in frequency for this value in the US than in other countries. The US scores high on the Hofstede dimensions of indulgence and individualism, which may correlate with hedonism. However, it should be noted that the UK also scores high on these dimensions and yet shows a less marked increase in hedonism.

Concerning *conformity* (observation 7), we would like to suggest that the COVID-19 pandemic has led to an increase of its importance in countries in which this value is traditionally less dominant. COVID-19 measures such as self-isolation and social distancing have challenged the acquiescence of populations. Therefore, the measures might have required more value change in countries with high scores for individualism in the Hofstede dimensions (Fig. 2.9). Our results seem to confirm this suggestion. The importance of conformity appears to have increased in countries with individualism scores above 50 (US, UK, South Africa) while remaining relatively stable for those with scores below 50 (Japan, South Korea, India).

Comparison with Existing Value Theories

We will now move to compare our results with what might be expected based on two prominent descriptive value theories, namely Schwartz' theory of universal values (Schwartz, 1992) and Inglehart's modernisation theory of value change (Inglehart, 2018). We start with the latter.

---

[13]Again we remind the reader that (changes in) frequencies cannot always be interpreted as (changes in) importance, there may be other reasons for changes in frequency. Perhaps, privacy and democracy are better guaranteed through laws and institutions in the other four countries, and this explains why they are less discussed. This is however speculative.

[14]We have tried to separate our value topic 'socio-economic equality' from the topic 'black lives matter', but that was not easy, and we might not have been fully successful. Apart from that, there is – as mentioned in the text- the possibility that BLM indirectly led to more attention to the value of 'socio-economic equality' in COVID news.

Inglehart (2018) has formulated two important value change hypotheses: the socialisation and scarcity hypothesis. The first holds that people's values are usually formed before adulthood and do not change much after that. The second states that virtually everyone values postmaterialist values like freedom and autonomy but prioritises materialist values like physical security and economic welfare under conditions of scarcity. Consequently, people's values reflect the conditions that were prevalent in the years before their adulthood. In increasingly affluent societies, one would therefore expect a gradual shift to postmaterialist values over time because the mix of generations in the total population changes over time. This general expectation indeed seems corroborated by empirical research (Inglehart, 2018). In addition, Inglehart allows for the possibility of more short-term value change due to crises or otherwise exceptional circumstances.

Lampert et al. (2021) report value changes due to COVID-19 pandemic based on surveys that used the methodology of the World Value Survey, which is based on Inglehart's theoretical work. It concerns changes in values between the first and fourth quarter of 2021 aggregated for 24 countries. As explained in detail in Appendix 5, we have translated these outcomes in terms of an increase or decrease of the values we considered in this study and compared them to the trends we found in our study: see Table 2.2 for the results.

In interpreting this result, two things are essential to keep in mind. First, the World Value Survey measures what we have called *personal* values, while our method is more geared up to measure *social* values. We defined social values above as "shared anchors that people use to justify their behaviour to others and to which they orient their actions to a greater or lesser degree." It should be noted that, understood in this way, social values are different from people's preferences or personal values, even if these are aggregated over the entire population.

Table 2.2 Comparison of value change found by Lampert et al. (2021) with our data (last two columns)

Value	Lampert et al. (2021)	Compared to pre-COVID-19	During COVID-19
Health and safety	+	++	−−
Economic welfare	−	+	+
Mental health	+	+	−
Socio-economic equality	+	−−	+
Freedom	+	−	+
Democracy	+	−−	+
Environmental sustainability	+	−	+
Privacy	+	−−	+
Hedonism	−	+	+
Conformity	−	−	+
Belonging	+	−	+
Overall fit (same direction)		2 out of 11	6 out of 11

Second, our method measures changes in the *frequency* of social values, while the World Value Survey measures changes in (subjective) *importance* of personal values. We cannot, therefore, directly compare our results with those from surveys like the World Value Survey. Nevertheless, we believe that one might expect that the trend we find in changes in the frequency of social value – i.e. whether a value is decreasing or increasing in frequency – may well correspond with changes in the subjective importance that people – individually as well as collectively – attribute to certain values. In other words, we may expect that if people subjectively value 'health' higher over time, we also see an increase in frequency of the social value of 'health' in newspaper articles. We may, therefore, expect *similar trends* even if we are not measuring the same construct.

In this light one striking observation is that the trend we observe *during* the Covid pandemic is similar to trend found by Lampert et al. (2021), while the trend we find compared to pre-COVID times seems opposite to the trend found by Lampert et al. (2021). This suggests that what Lampert et al. are actually measuring is a value change *during* the covid pandemic, instead of value change *due to* COVID. This is also not unlikely because they characterise their first measurements in early 2020 as "*just* before the pandemic hit most countries early in 2020" (p. 2; emphasis added).[15] However, our data strongly suggest that value change already took place during the first quarter of 2021 (see, e.g. Fig. 2.7), so that doubts may be raised about their claim that their Q1 survey data really measure pre-covid conditions.

It is also interesting to compare the trends of value change we find with Schwartz theory of basic values, which we briefly described in Sect. 2.2. Figure 2.12 shows the ten basic Schwartz values; Schwartz takes values close together in this figure to be (motivationally) reinforcing, while values far apart or opposite are assumed to be (motivationally) opposite or contradictory. This implies that if, for example, security values become more important, self-direction, and universalism will be emphasised less.

We may use this theoretical idea to formulate certain hypotheses about how values will change. To do so, we have associated our list of values with the Schwartz values. In addition, we might assume that a pandemic like the COVID-19 one will lead to more emphasis on security values (Steinert, 2020). However, our data suggests that the direction of value change for security and survival values reverses during the pandemic. Therefore, we have assumed that the changes in other values are a function of the change in the value of safety and health, following Schwartz' logic of values that reinforce each other or are opposite. The results are shown in Table 2.3. Overall, we find a relatively good fit with what one would expect based on Schwartz' value theory. Particularly for his universalism and self-direction values, we witness a good fit with our observations (see Table 2.3). Therefore, our

---

[15] The interviews were done online between 23 January and 11 March 2020 (Lampert et al., 2021:45).

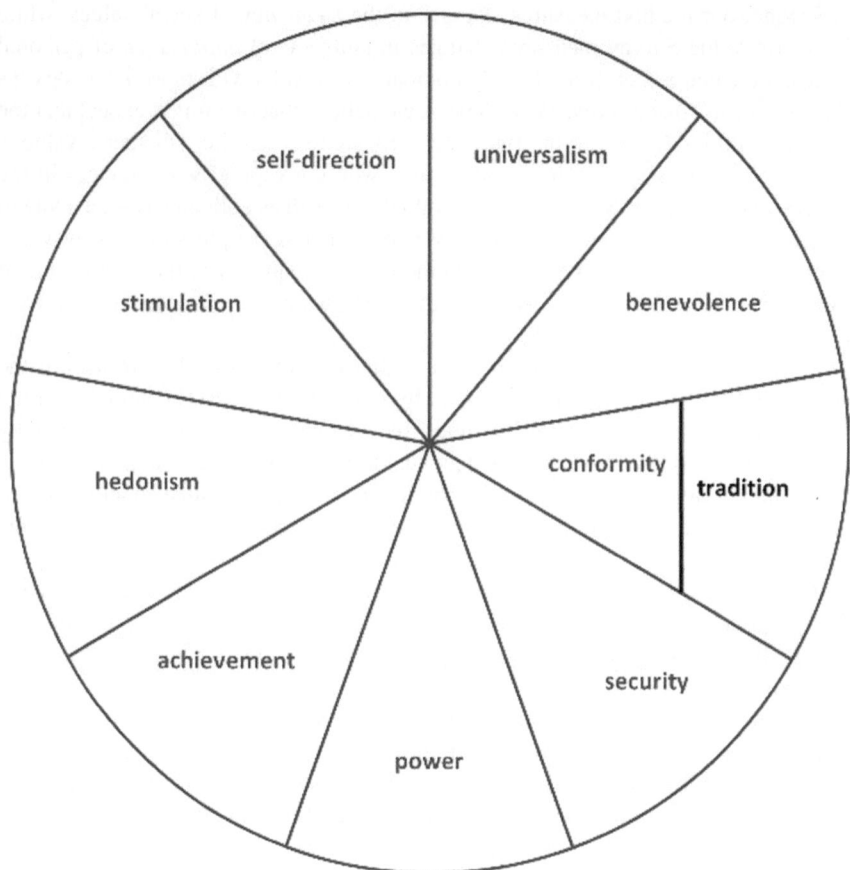

Fig. 2.12 Schwartz values

observations seem to confirm Schwartz's idea that changes in different values are related to each other.

Moral Implications

What do our findings tell us about moral values? While we did not directly trace moral values or changes in them, one might argue that values in news articles reflect values that are considered morally important in a society or country. They may at least reflect what the writers of such news think that people consider (or should consider) morally relevant values. This is agnostic on whether we also always have moral reasons to consider such values important.

Still, some of our findings may have indirect moral implications. One of these implications is related to how we can best phrase some of the moral issues raised by the COVID-19 pandemic. One popular phrase to describe this is "moral dilemmas". For example, it has often been suggested that we need to choose between "life" versus "livelihood", or between the values of "safety and health" versus the value of

Table 2.3 Comparison with value change expectations based on Schwartz' value theory

Value	Schwartz value	Compared to pre-covid		During covid	
		Expectation	This study	Expectation	This study
Health and safety	security	+	++	−	−−
Economic welfare	security	+	+	−	+
Mental health		?	+	?	−
Socio-economic equality	universalism	−	−−	+	+
Freedom	self-direction	−	−	+	+
Democracy	self-direction/ universalism	−	−−	+	+
Environmental sustainability	Universalism	−	−	+	+
Privacy	self-direction	−	−−	+	+
Hedonism	Hedonism	−	+	+	+
Conformity	conformity	+	−	−	+
Belonging	security	+	−	−	+
Overall fit (same direction)		7 out of 10		7 out of 10	

"economic welfare" in deciding on measures against the virus (e.g. Sharma & Mahendru, 2020; Fernandez et al., 2021).[16] Some have also voiced the fear that the COVID-19 pandemic comes at the costs of (moral) values we hold dear like democracy, freedom and privacy.[17]

Without denying the possibility of dilemmas and trade-offs, we find no support for the idea that the moral questions concerning COVID-19 should be understood in the form of dilemmas. For example, while we see a decline in values like democracy, freedom, and privacy at the start of the crisis, their frequency goes up later, without necessarily rising at the expense of attention for safety and health. Furthermore, we consider it to be telling that the same trend can be observed as even more pronounced in low-income countries from the Global South.

Similarly, there seems to be little evidence that we face a dilemmatic choice between "safety and health" versus "economic welfare". The countries in our sample in which we see a relatively strong emphasis on economic welfare – Japan and South Korea – are also those that do best at minimising the effects of the Covid pandemic in terms of health and fatalities (see Appendix 3). Moreover, the overall stringency of measures in these two countries was not larger, or even smaller, than in the other four countries. This may be due to the fact that these countries have taken measures earlier (see Appendix 4). Furthermore, data from *Our World in Data*

---

[16] For another example, see https://www.lse.ac.uk/philosophy/blog/2020/10/21/lives-v-livelihoods/. Accessed 9 July 2021.

[17] For example, https://freedomhouse.org/report/special-report/2020/democracy-under-lockdown. Accessed 9 July 2021.

suggest a significant *positive* correlation between the degree to which countries succeeded in reducing new cases and fatalities and how successful they were in reducing the negative economic effects of the crisis (Ritchie et al., 2020).[18] Therefore, the suggestion that we face a moral dilemma, in the terms we have set out her, seems misleading (at best) and morally dangerous (at worst). This is because it stands to be misused by policy makers to pursue favoured policies for which there is no firm moral ground. Of course, it does not follow that other pandemic-related choices faced by governments in the future will not be dilemmatic; this will very much depend on the specific case. Whether a choice is dilemmatic is something we may sometimes only find out along the way, and it might not be evident at the moment of choice.

More generally, our observations may offer ground for some optimism in the sense that after a worrying decrease in the frequency of some morally important values like democracy and socio-economic equality, we clearly see these values increase in frequency at a later stage. We might interpret this as a sign of what may be termed *moral resilience*, i.e. the ability of a society to pay attention to morally important values despite these values being put under pressure in a crisis. That does not necessarily mean that these values are also better addressed or realised. We cited literature in Sect. 2.2 that gives reason to doubt so, but this observation at least implies that these values get more attention in the news and are connected to collective discussions about the pandemic. That is at least a start to ensuring that these moral values get the attention they deserve. One development that is nevertheless worrying in this respect is that the perceived importance to the value of mental health seems to have declined during the crisis (see Figs. 2.1, 2.2, 2.3, 2.4, 2.5, and 2.6), as this value certainly seems under pressure and would seem to require more rather than less attention from a moral point of view; this may then be considered an important moral blind spot.

2.6 Conclusions

We find that the first few months of the COVID-19 pandemic led to a punctuated change in social values. While the value of safety and health sharply increased in frequency, the values of *democracy*, *privacy* and *socio-economic equality* significantly declined. However, after this first shock, we see a relative decline in the value of *safety and health* in COVID-related news, while most other values have increased in frequency. While we lack the data to make strong claims about long-term effects, the pattern we find suggests that it may well be possible that the long-term effects of the pandemic in terms of social value change are limited.

---

[18] https://ourworldindata.org/covid-health-economy, Accessed 1 July 2021. Whether this correlation is the same for the remainder of the pandemic remains to be seen, of course.

Perhaps somewhat surprisingly, we find that the three seemingly contradicting studies we mentioned in the introduction are all right in some respect: Reeskens et al. (2021) are correct that the long-term effect of the pandemic on values may be limited; Steinert (2020) is right in the sense that the pandemic at least initially led to more stress on security and survival values; and the apparent change toward postmaterialist values found by Lampert et al. (2021) may well reflect value changes during the pandemic rather than a value change compared to pre-covid times. We further conclude that the patterns of value change we found are more or less in line with Schwartz' value theory that poses that specific values have opposing tendencies.

We also found and discussed some differences in value change between countries, which we could – to some extent – explain by economic and cultural differences between those countries. Concerning moral implications, we found no evidence that the pandemic has a clearly dilemmatic character. Instead, our findings suggest that the countries studied showed some moral resilience in the sense that morally important values began to increase in frequency again after their initial decline. While this certainly does not mean that these values are sufficiently addressed in actual policies, it means that they are at least addressed in news articles.

Acknowledgement   This publication is part of the project ValueChange that has received funding from the European Research Council (ERC) under the European Union's Horizon 2020 research and innovation programme under grant agreement No 788321. It also contributes to the research programme Ethics of Socially Disruptive Technologies, which is funded through the Gravitation programme of the Dutch Ministry of Education, Culture, and Science and the Netherlands Organization for Scientific Research (NWO grant number 024.004.031).

Appendix 1: Text Corpora Used

Countries	Newspapers	No. of articles	Dataset
India	Hindustan Times, The Indian Express	47,342	Ghasiya and Okamura (2021)
Japan	The Japan Times, Asahi Shimbun, Mainichi Shimbun	21,039	Ghasiya and Okamura (2021)
South Korea	Korea Herald, Korea Times	102,278	Ghasiya and Okamura (2021)
UK	bbc.co.uk, mirror.co.uk, sky.com, express.co.uk, theguardian.com, thesun.co.uk, metro.co.uk, dailymail.co.uk, thetimes.co.uk, cnet.com, msn.com, alaraby.co.uk, skysports.com, dailystar.co.uk, thomsonreuters.com, digitalspy.com, channel4.com, parliament.uk, www.gov.uk, hitc.com, reuters.com, telegraph.co.uk, economist.com, nature.com, bmj.com, www.nhs.uk, ft.com, ox.ac.uk, barclays.co.uk, europa.eu	5000	Aylien Ltd. (2020)

Countries	Newspapers	No. of articles	Dataset
USA	thehill.com, washingtonpost.com, cbslocal.com, chicagotribune.com, theadvocate.com, qz.com, businessinsider.com, nypost.com, rollingstone.com, huffingtonpost.com, cnbc.com, forbes.com, deadline.com, cnn.com, sfgate.com, nbcnews.com, go.com, denverpost.com, politico.com, breitbart.com, foxnews.com, psu.edu, msn.com, ucdavis.edu, bgr.com, npr.org, bizjournals.com, nydailynews.com, latimes.com, google.com, cnet.com, nbcsports.com, usatoday.com, newsweek.com, brobible.com, motorsport.com, usnews.com, marketwatch.com, thedailybeast.com, cbsnews.com, bustle.com, dailycaller.com, cbssports.com, yahoo.com, psychologytoday.com, mashable.com, buzzfeed.com, vox.com, nymag.com, delta.com, complex.com, scientificamerican.com, techcrunch.com, hbr.org, fastcompany.com, foxbusiness.com, vanityfair.com, androidcentral.com, pbs.org, cdc.gov, ca.gov, wired.com, newyorker.com, aol.com, fivethirtyeight.com, apnews.com, gsmarena.com, slate.com, variety.com, billboard.com, snopes.com, theatlantic.com, pitchfork.com, tmz.com, harvard.edu, nih.gov, cosmopolitan.com, bloomberg.com, acs.org, issuu.com, sciencedaily.com, cisco.com, ew.com, techtarget.com, eonline.com, chron.com, menshealth.com, legacy.com, vulture.com, nba.com, digitaltrends.com, yelp.com, mit.edu, producthunt.com, zdnet.com, umich.edu, archdaily.com, arizona.edu, nytimes.com, usda.gov	5000	Aylien Ltd. (2020)
South-Africa	news24.com	4296	Aylien Ltd. (2020)
Worldwide	Reuters	91,180	Thompson (2020)

Appendix 2: Values Identified

The first eight values in the table below are based on a brainstorm of the authors. The last three values were added on basis of the values that resulted from the brainstorm with the values found in Liscio et al. (2022); they latter let two teams of human annotators identify values in text corpora based on a PVE (Participatory Value Evaluation) study on relaxing COVID-19 measures in the Netherlands (Mouter et al. 2021).

Value	Corresponding values from Liscio et al. (2021)
Health and safety	Safety and health, safety, control
Economic welfare	Economic security, economic prosperity, feasibility
Mental health	Mental health, well-being, care
Socio-economic equality	Equality, fairness
Freedom	Autonomy
Democracy	
Environmental sustainability	
Privacy	
Hedonism	Pleasure, enjoyment, being social
Conformity	Acceptance of misbehaviour, conformity
Belonging	Belonging to a group, nuclear family

Appendix 3: COVID-19 Cases and Deaths over Time

The graphs in this appendix are based on data from Ritchie et al. (2020). Retrieved from https://ourworldindata.org/coronavirus on 16 June 2021. The blue line indicates the number of new cases per million inhabitants in a country on a daily base (left axis), while the orange line indicates the number of new deaths per million inhabitants on a daily base (right axis).

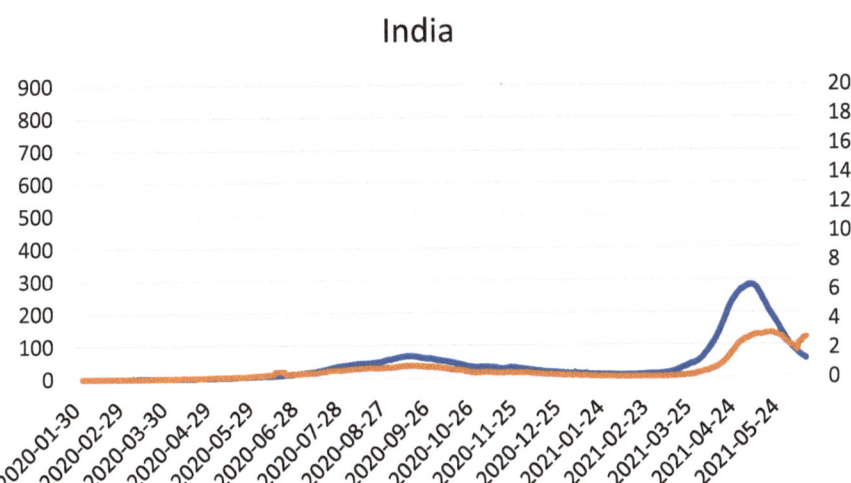

India

South Africa

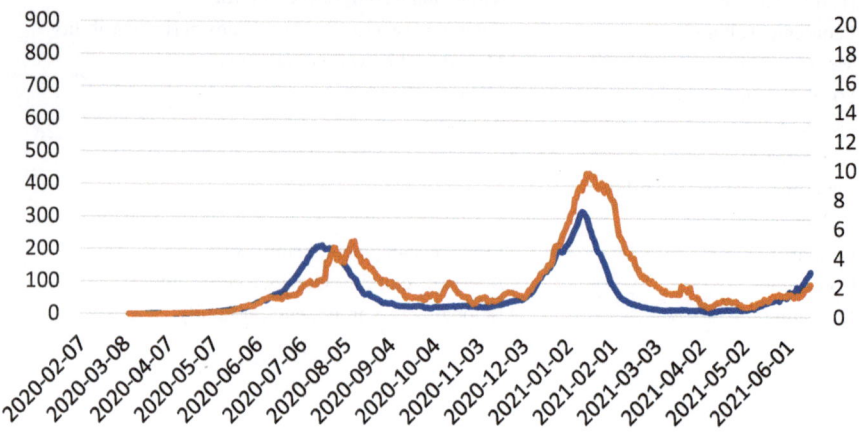

UK

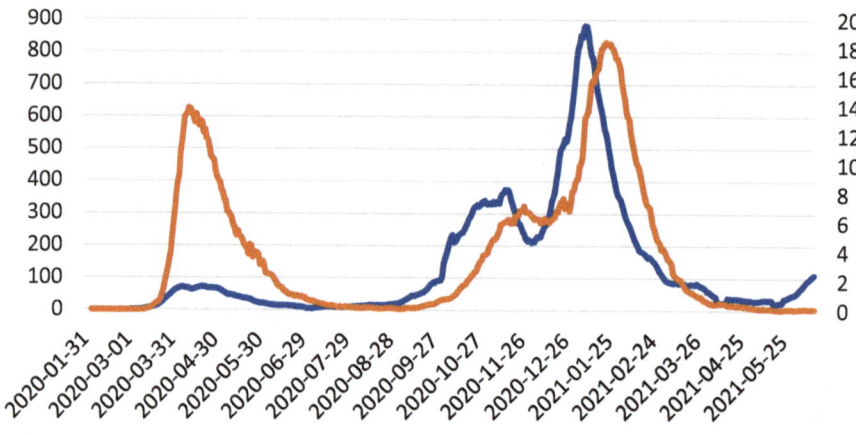

US

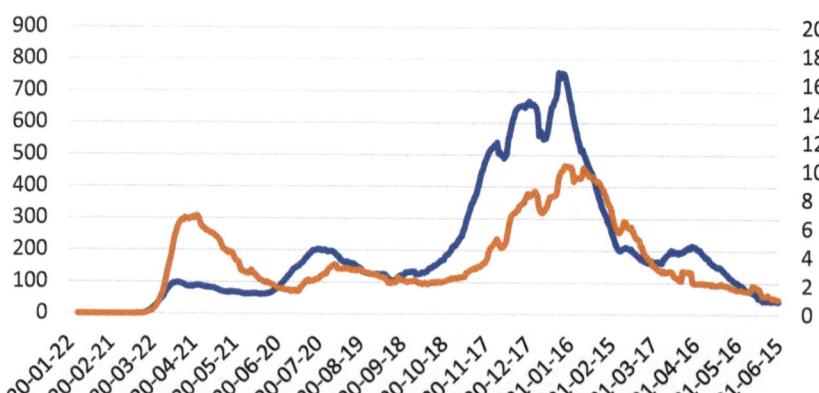

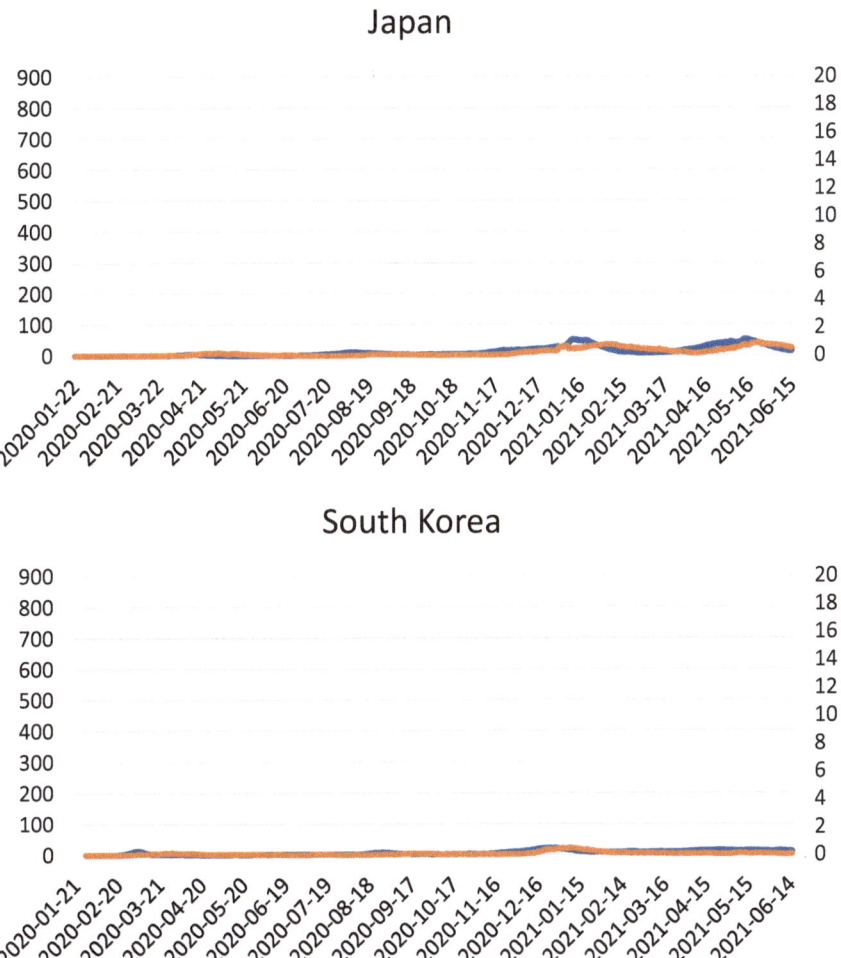

Appendix 4: Stringency of Measures

The graphs in this appendix are based on data from Ritchie et al. (2020). Retrieved from https://ourworldindata.org/covid-stringency-index on 16 June 2021. These data are originally from Hale et al. (2021). The following explanation is given at the website about the used stringency index: "The Oxford Coronavirus Government Response Tracker (OxCGRT) project calculate a Stringency Index, a composite measure of nine of the response metrics. The nine metrics used to calculate the Stringency Index are: school closures; workplace closures; cancellation of public events; restrictions on public gatherings; closures of public transport; stay-at-home requirements; public information campaigns; restrictions on internal movements;

and international travel controls. ... The index on any given day is calculated as the mean score of the nine metrics, each taking a value between 0 and 100" (https://ourworldindata.org/covid-stringency-index, accessed 8 July 2021).

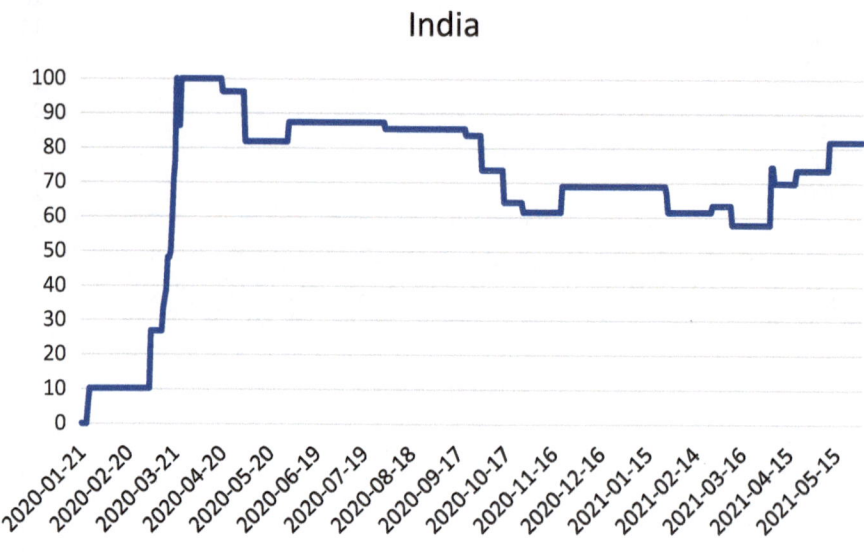

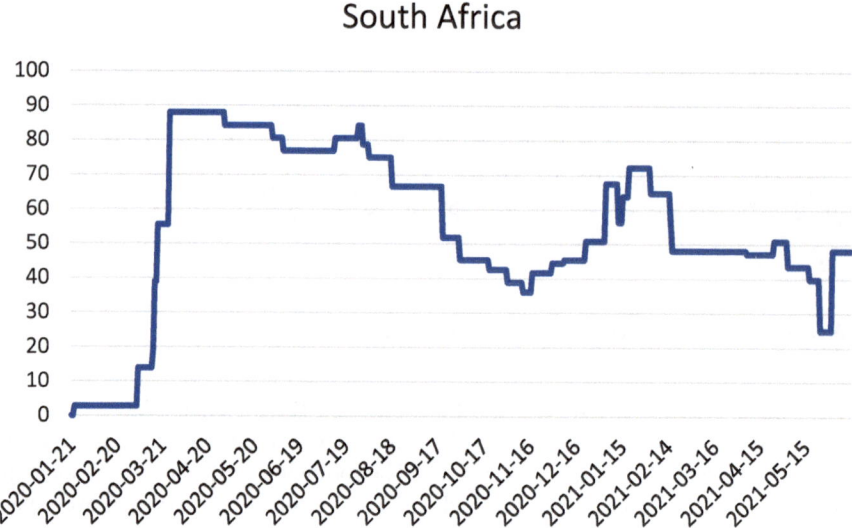

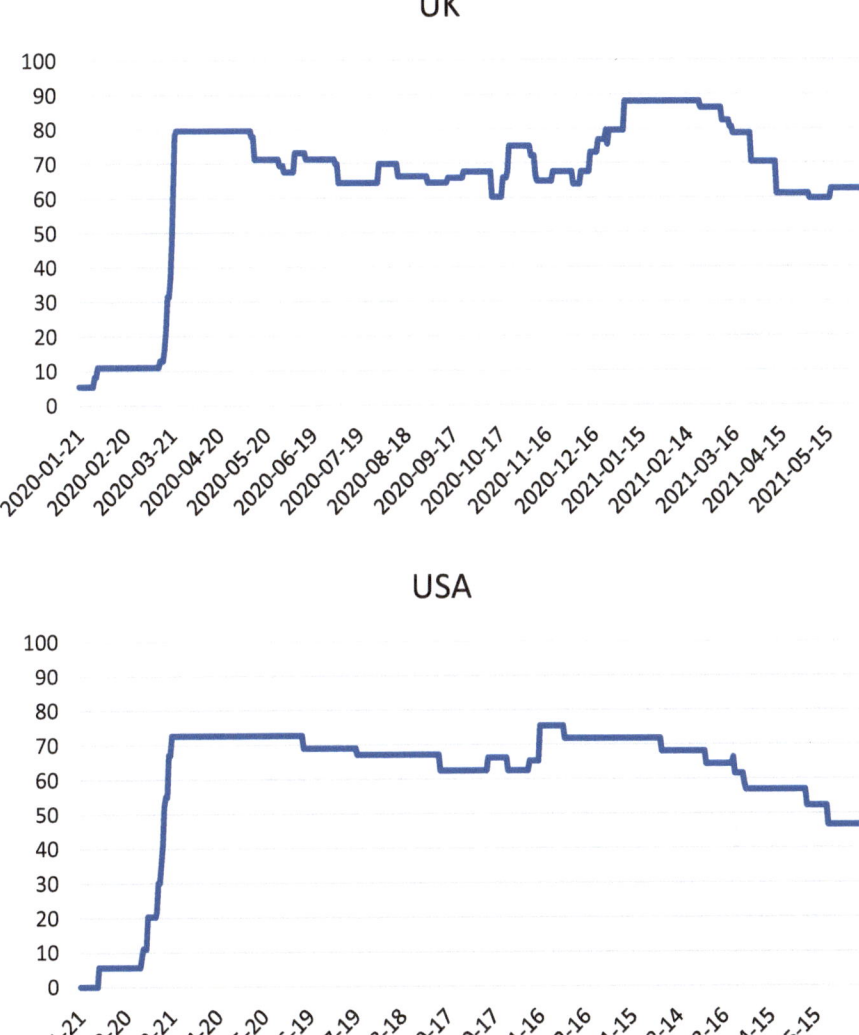

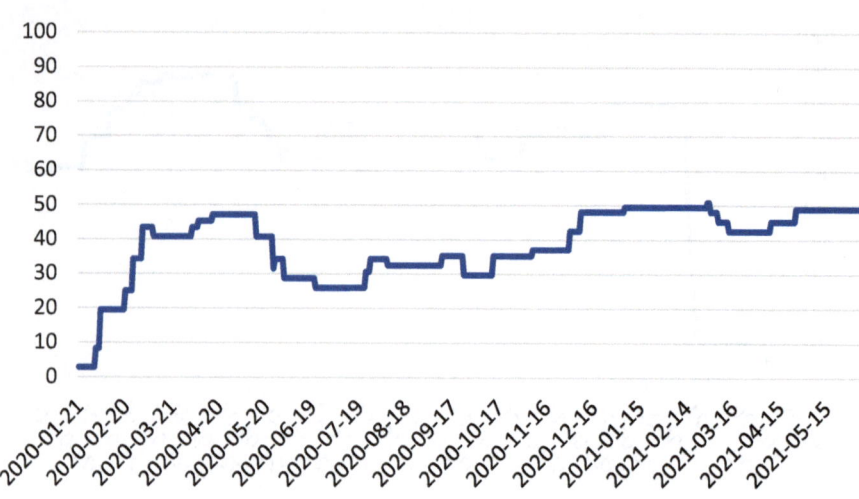

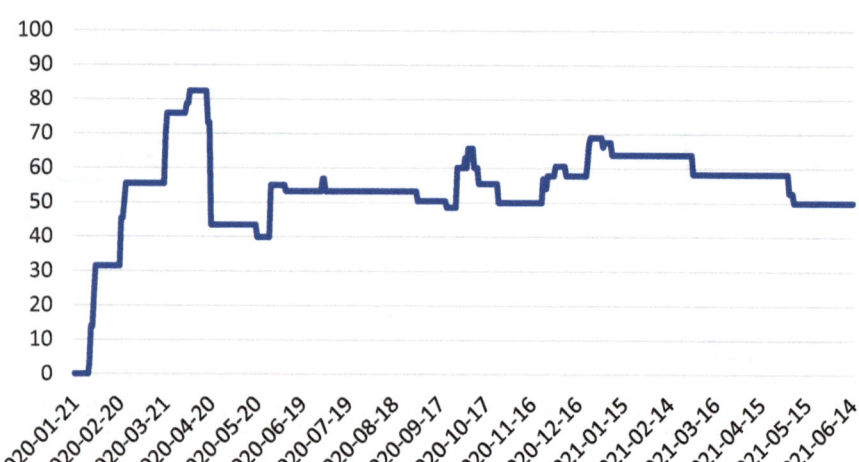

Appendix 5: Value Change Found in World Value Survey

The data in this appendix are based on a value survey done by Lampert et al. (2021).
We have associated the relevant items in their survey with values in our study
(according to our interpretation), and we have then looked whether they observe a
decrease or increase in the importance of these values if we assume that the items
measured are indicative for the values with which we associated them. The last
column indicates the direction of change of each of the values, we derived from
this survey.

Value	Value survey 24 countries (Lampert et al. 2021)			
	Item	Q1–2020	Q4–2020	Direction of change
Health and safety	In order to prevent any risks, I take precautionary measures	3.74	3.77	Increase
Economic welfare	Materialism/postmaterialism index (higher = postmaterialism)	2.33	2.43	Decrease
Mental health	I often feel lonely	2.98	2.9	Increase
	I sometimes feel that the future holds nothing for me	2.99	3.08	
	I feel let down by society	2.98	3.07	
	Life is easy	2.49	2.45	
Socio-economic equality	Every person in the world should be treated equally	4.14	4.2	Increase
	I think that differences between high and low incomes should be smaller	3.98	4.04	
Freedom	Control/freedom index (higher = more freedom)	3.47	3.52	Increase
	Materialism/postmaterialism index (higher = postmaterialism)	2.33	2.43	
Democracy	The country really needs more law and order and not more civil rights	3.04	2.93	Increase
	Materialism/postmaterialism index (higher = postmaterialism)	2.33	2.43	
Environmental sustainability	I worry about the damage humans cause to the planet	4.06	4.12	Increase
	I try living eco-consciously	3.82	3.86	
Privacy	Control/freedom index (higher = more freedom)	3.47	3.52	Increase
Hedonism	I often have the urge to experience something new	3.71	3.64	Decrease
	My most important aims are to have fun and enjoy myself	3.55	3.36	

Value	Value survey 24 countries (Lampert et al. 2021)			
	Item	Q1–2020	Q4–2020	Direction of change
Conformity	Control/freedom index (higher = more freedom)	3.47	3.52	Decrease
	Etiquette (rules determining what good manners are) is very important to me	3.84	3.8	
	Materialism/postmaterialism index (higher = postmaterialism)	2.33	2.43	
	The country really needs more law and order and not more civil rights	3.04	2.93	
Belonging	I feel strongly involved with what is happening in my community	3.28	3.31	Increase

References

Aylien Ltd. (2020). *Coronavirus news dataset*. https://aylien.com/resources/datasets/coronavirus-dataset?utm_referrer=https%3A%2F%2Faylien.com%2F

Bhat, S. A., Bashir, O., Bilal, M., Ishaq, A., Din Dar, M. U., Kumar, R., Bhat, R. A., & Sher, F. (2021). Impact of COVID-related lockdowns on environmental and climate change scenarios. *Environmental Research, 195*, 110839. https://doi.org/10.1016/j.envres.2021.110839

Blei, D. M., & Lafferty, J. D. (2009). Topic models. In A. Srivastava & M. Sahami (Eds.), *Text mining: Theory and applications* (pp. 71–89). Taylor and Francis. https://doi.org/10.1145/1143844.1143859

Chan, E. Y., & Saqib, N. U. (2021). Privacy concerns can explain unwillingness to download and use contact tracing apps when COVID-19 concerns are high. *Computers in Human Behavior, 119*, 106718. https://doi.org/10.1016/j.chb.2021.106718

Cheval, S., Mihai Adamescu, C., Georgiadis, T., Herrnegger, M., Piticar, A., & Legates, D. R. (2020). Observed and potential impacts of the COVID-19 pandemic on the environment. *International Journal of Environmental Research and Public Health, 17*(11), 4140. https://doi.org/10.3390/ijerph17114140

Cifuentes, M. P., Rodriguez-Villamizar, L. A., Rojas-Botero, M. L., Alvarez-Moreno, C. A., & Fernández-Niño, J. A. (2021). Socioeconomic inequalities associated with mortality for COVID-19 in Colombia: A cohort nationwide study. *Journal of Epidemiology and Community Health, 75*, 610–615. https://doi.org/10.1136/jech-2020-216275

Clouston, S. A. P., Natale, G., & Link, B. G. (2021). Socio-economic inequalities in the spread of coronavirus-19 in the United States: A examination of the emergence of social inequalities. *Social Science & Medicine, 268*, 113554. https://doi.org/10.1016/j.socscimed.2020.113554

de Wildt, T. E., van de Poel, I. R., & Chappin, E. J. L. (2021). Tracing long-term value change in (energy) technologies: Opportunities of probabilistic topic models using large data sets. *Science, Technology, & Human Values, 47*(3), 429–458. https://doi.org/10.1177/01622439211054439

de Wildt, T. E., Chappin, E. J., van de Kaa, G., & Herder, P. M. (2018). A comprehensive approach to reviewing latent topics addressed by literature across multiple disciplines. *Applied Energy, 228*, 2111–2128. https://doi.org/10.1016/j.apenergy.2018.06.082

Demski, C., Butler, C., Parkhill, K. A., Spence, A., & Pidgeon, N. F. (2015). Public values for energy system change. *Global Environmental Change, 34*, 59–69. https://doi.org/10.1016/j.gloenvcha.2015.06.014

Fernandez, B., Marcela, P., Mejia-Mantilla, C., Olivieri, S. D., Ibarra, G. L., Haaker, R., & Javier, F. (2021). *Lives or livelihoods? The costs of staying healthy – COVID-19 in LAC Washington.* World Bank.

Gerke, S., Shachar, C., Chai, P. R., & Glenn Cohen, I. (2020). Regulatory, safety, and privacy concerns of home monitoring technologies during COVID-19. *Nature Medicine, 26*(8), 1176–1182. https://doi.org/10.1038/s41591-020-0994-1

Ghasiya, P., & Okamura, K. (2021). Investigating COVID-19 news across four nations: A topic modeling and sentiment analysis approach. *IEEE Access, 9,* 36645–36656. https://doi.org/10.1109/ACCESS.2021.3062875

Hale, T., Angrist, N., Goldszmidt, R., Kira, B., Petherick, A., Phillips, T., Webster, S., Cameron-Blake, E., Hallas, L., Majumdar, S., & Tatlow, H. (2021). A global panel database of pandemic policies (Oxford COVID-19 Government Response Tracker). *Nature Human Behaviour, 5*(4), 529–538. https://doi.org/10.1038/s41562-021-01079-8

Hofstede, G. H., Hofstede, G. J., & Minkov, M. (2010). *Cultures and organisations: Software of the mind : Intercultural cooperation and its importance for survival* (3rd ed.). McGraw-Hill.

Inglehart, R. (2018). *Cultural evolution: People's motivations are changing, and reshaping the world, American political thought.* Cambridge University Press.

Inglehart, R., & Welzel, C. (2009). How development leads to democracy: What we know about modernization. *Foreign Affairs, 88*(2), 33–48.

Jasanoff, S., & Kim, S.-H. (2015). *Dreamscapes of modernity: Sociotechnical imaginaries and the fabrication of power.* The University of Chicago Press.

Kauffman, C. M. (2021). Democratization. In *Encyclopedia britannica.* https://www.britannica.com/topic/democratization

Lampert, M., Inglehart, R., Metaal, S., Schoemaker, H., & Papadongonas, P. (2021). Two faces of Covid-19 impact: The pandemic ignites fear, but boosts progressive ideals and calls for inclusive economic growth. *Measuring the pandemic's impact on social values, emotions and priorities in 24 countries.* Retrieved from https://glocalities.com/latest/reports/valuestrends: Glocalities.

Liscio, E., van der Meer, M., Jonker, C. M., & Murukannaiah, P. K. (2021). *Axies: Identifying and evaluating context-specific values.* Proceedings of the 20th International Conference on Autonomous Agents and MultiAgent Systems, Virtual Event, United Kingdom.

Liscio, E., Dondera, A., Geadau, A., Jonker, C., & Murukannaiah, P. (2022). *Cross-domain classification of moral values.* Findings of the Association for Computational Linguistics: NAACL 2022.

Lopez, L., Hart, L. H., & Katz, M. H. (2021). Racial and ethnic health disparities related to COVID-19. *JAMA, 325*(8), 719–720. https://doi.org/10.1001/jama.2020.26443

Mouter, N., Hernandez, J. I., & Itten, A. V. (2021). Public participation in crisis policymaking. How 30,000 Dutch citizens advised their government on relaxing COVID-19 lockdown measures. *PLoS One, 16*(5), e0250614. https://doi.org/10.1371/journal.pone.0250614

Perry, B. L., Aronson, B., & Pescosolido, B. A. (2021). Pandemic precarity: COVID-19 is exposing and exacerbating inequalities in the American heartland. *Proceedings of the National Academy of Sciences, 118*(8), e2020685118. https://doi.org/10.1073/pnas.2020685118

Rawls, J. (1999 [1971]). *A theory of justice* (Revised Edition ed.). The Belknap Press of Harvard University Press.

Reeskens, T., Muis, Q., Sieben, I., Vandecasteele, L., Luijkx, R., & Halman, L. (2021). Stability or change of public opinion and values during the coronavirus crisis? Exploring Dutch longitudinal panel data. *European Societies, 23*(sup1), S153–S171. https://doi.org/10.1080/14616696.2020.1821075

Reicher, S., & Stott, C. (2020). On order and disorder during the COVID-19 pandemic. *British Journal of Social Psychology, 59*(3), 694–702. https://doi.org/10.1111/bjso.12398

Repucci, S., & Slipowitz, A. (2020). *Democracy under lockdown: The impact of COVID-19 on the global struggle for freedom.* Freedom House. Retrieved from https://freedomhouse.org/report/special-report/2020/democracy-under-lockdown

Ritchie, H., Mathieu, E., Rodés-Guirao, L., Appel, C., Giattino, C., Ortiz-Ospina, E., ... & Roser, M. (2020). Coronavirus pandemic (COVID-19). *Our World in Data* 2021. Published online at OurWorldInData.org. Accessed 16 June 2021.

Rokeach, M. (1973). *The nature of human values*. The Free Press.

Rume, T., & Didar-Ul Islam, S. M. (2020). Environmental effects of COVID-19 pandemic and potential strategies of sustainability. *Heliyon, 6*(9), e04965. https://doi.org/10.1016/j.heliyon.2020.e04965

Rupani, P. F., Nilashi, M., Abumalloh, R. A., Asadi, S., Samad, S., & Wang, S. (2020). Coronavirus pandemic (COVID-19) and its natural environmental impacts. *International Journal of Environmental Science and Technology, 17*, 4655–4666. https://doi.org/10.1007/s13762-020-02910-x

Schwartz, S. H. (1992). Universals in the content and structure of values: Theoretical advances and empirical tests in 20 countries. In M. P. Zanna (Ed.), *Advances in experimental social psychology* (pp. 1–65). Academic.

Sharma, G. D., & Mahendru, M. (2020). Lives or livelihood: Insights from locked-down India due to COVID19. *Social Sciences & Humanities Open, 2*(1), 100036. https://doi.org/10.1016/j.ssaho.2020.100036

Steg, L., & De Groot, J. I. M. (2012). Environmental values. In S. Clayton (Ed.), *The Oxford handbook of environmental and conservation psychology* (pp. 81–92). Oxford University Press.

Steinert, S. (2020). Corona and value change. The role of social media and emotional contagion. *Ethics and Information Technology, 23*(1), 59–68. https://doi.org/10.1007/s10676-020-09545-z

Sun, M., Zhang, H., Zhao, Y., & Shang, J. (2014). From post to values: Mining Schwartz values of individuals from social media. *Communications in Computer and Information Science, 489*, 206–219. https://doi.org/10.1007/978-3-662-45558-6

Szmigiera, M. (2021). *Impact of the coronavirus pandemic on the global economy – Statistics & facts*. Last Modified 7–5, Accessed 22–5. https://www.statista.com/topics/6139/covid-19-impact-on-the-global-economy/#dossierSummary

Thompson, A. (2020). 2.7 million news articles and essays. *Reuters*. https://components.one/datasets/all-the-news-2-news-articles-dataset

World Health Organization. (2006). *Constitution of the world health organization – Basic documents, supplement.*

Open Access   This chapter is licensed under the terms of the Creative Commons Attribution 4.0 International License (http://creativecommons.org/licenses/by/4.0/), which permits use, sharing, adaptation, distribution and reproduction in any medium or format, as long as you give appropriate credit to the original author(s) and the source, provide a link to the Creative Commons license and indicate if changes were made.

The images or other third party material in this chapter are included in the chapter's Creative Commons license, unless indicated otherwise in a credit line to the material. If material is not included in the chapter's Creative Commons license and your intended use is not permitted by statutory regulation or exceeds the permitted use, you will need to obtain permission directly from the copyright holder.

Chapter 3
What Has COVID-19 Taught Us About Democracy? Relational Democracy and Digital Surveillance Technologies

Elena Ziliotti

3.1 Introduction

The COVID-19 pandemic has called attention to the absence in the West of institutions and procedures to debate health surveillance tools in a democratic way. The democratic ideal of governance entails that the political decisions that have the potential to affect the members of society are taken by the citizens, or indirectly by their representatives. The democratic principle of self-governance suggests that in a democratic society, the decision to experiment with digital surveillance technologies must not be insulated from public debates, especially when such experimentation (or the abstention from it) can significantly affect parts of the public in either the short- or long-term. Inputs from different parts of the public must inform decisions on the principles that guide digital surveillance technologies and the use of these technologies must be the subject of debate by the citizens or their representatives.

The democratic idea of self-government contrasts with the practice of experimentation with digital surveillance technologies in several Western democracies. Private corporations like Facebook and Google exercise unlimited power on algorithms that structure the national and international digital public sphere of many

E. Ziliotti (✉)
Delft University of Technology, Delft, The Netherlands
e-mail: e.ziliotti@tudelft.nl

© The Author(s) 2022
M. J. Dennis et al. (eds.), *Values for a Post-Pandemic Future*, Philosophy
of Engineering and Technology 40, https://doi.org/10.1007/978-3-031-08424-9_3

democratic societies (Simons and Gosh 2020: 2).[1] Yet, during the COVID-19 crisis, many Western democratically-elected governments decided to refrain from using certain new technologies (such as control of phone geolocation data, transportation card data, and CCVTs footage) to manage the pandemic.[2] An exception to this is the introduction of voluntary contact-tracing apps to track the COVID-19 cases, which were introduced in some Western countries with the promise that they could be useful in preventing the spread of COVID-19, while also respecting citizens' privacy. However, the introduction of these apps proved not to be well-timed and few of these apps turned out to be effective.[3]

This chapter contributes to the existing literature on the relationship between democracy and digital surveillance. It argues that the paradoxical relation of Western countries with capitalist and state-driven forms of experimentations with digital surveillance technologies urges us to rethink how democracies must engage with these technological experimentations. I argue that the relational conception of democracy offers a viable approach to explain how digital surveillance can be put under democratic control and how such experimentations with new digital surveillance technologies can take place. The relational account of democracy offers a context-sensitive approach to digital surveillance technologies that places public deliberation at the centre of the democratic decision-making process. Furthermore, reaching a decision independently from the public discussion can deprive decision making from key contextual information and the epistemic inputs of different members of society that may hold the key to finding a successful solution in a given situation. Not only does this suggest that the existence of digital surveillance activities conducted by private parties should be a matter of democratic deliberation, but also that, if digital surveillance technologies are going to have a strong societal impact on the fight against COVID-19, then there should be a public discussion on the topic. To illustrate how relational democratic experimentations with digital surveillance technologies would look like, I will discuss the case of South Korea's experimentation with new digital surveillance technologies during 2020, the first year of the pandemic. This chapter does not argue for an increase in surveillance, nor does it support forms of state

---

[1] The use of digital surveillance technologies managed by self-interested parties is the subject on a heated debate in Wester democracies. Shoshana Zuboff maintains that the deploy of digital surveillance technologies by private corporations has led to a new form of capitalism that exploits users' private experience for the production of data to be sold into the market (e.g. targeted advertising) (Zuboff, 2019a, 2019b). Other scholars like, Stephen Graham and David Wood argue that the adoption of digital surveillance can worsen the position of already marginalized groups (Graham & Wood, 2003). For other critical analyses of digital surveillance, see Gilliom (2001) and Bogard (1996).

[2] Exception to these trend are South and North Dakota. In these states, a contract tracing app gathers citizens' last 10-day location data through GPS, Wi-Fi and cell towers (North Dakota Health, 2020; State of South Dakota, 2020).

[3] The slow progress of contact tracing apps has multiple reasons. In the USA, seven months into the COVID outbreak, these apps were "hampered by sluggish and uncoordinated development, distrust of technology companies, and inadequate advertising budgets and messaging campaigns" (De La Garza, 2020). Low numbers of downloads seem to be a key issue also for the ineffectiveness of the NHS's contact tracing app *COVID app* in UK, where only 28% of the population downloaded the app (Lewis, 2021).

digital surveillance against capitalist digital surveillance. The chapter argues that in a democratic society, legitimate and epistemically-superior experimentation with digital surveillance must be democratically controlled.

Drawing on pragmatist ethics and political discussions on relational democracy, Sect. 3.2 introduces the relational democratic approach to digital surveillance technologies. The interdependence between the individual and the community is at the basis of relational accounts of democracy, which also underlies a specific relationship between democracy and new technologies (Sect. 3.3). To clarify how this different democratic approach to technologies could look like in practice, in Sect. 3.4 I discuss the case of Korea's experimentation with new digital surveillance technologies in their battle against COVID-19.

3.2   Relational Ideal of Democracy

The relational conception of democracy considers democracy as 'a way of life' or 'a culture'; it emphasizes the social and collective experience of democratic life. Elizabeth Anderson maintains that democracy can be understood on three levels: as a mode of governance, as a membership organization and also as 'a way of life' (Anderson, 2009); each of these levels interacts with the others. Unlike liberal democrats, who understand democracy more in institutional terms, relational democrats emphasize the kind of social relations that tight fellow citizens together in a democratic society.[4]

The relational conception of democracy does not reduce democracy to a 'communitarian' ideal, nor does it deny the importance of the cultivation of a person's individuality. It maintains that such cultivation is a social phenomenon and individuality can be developed only through social relationships with others. As John Dewey, the first philosopher to formulate a relational conception of democracy puts it, democracy is "the greatest experiment of humanity – that of living together in ways in which the life of each of us is profitable in the deepest sense of the word, *profitable both to a single person and helpful in the building up of the individuality of others*" (my emphasis, Dewey, 1938/1991: 303). In Dewey's terms and those relational democratic theorists that followed him, the goal of a democratic community is the personal development of its members, but this is not an individual enterprise as individuality can be achieved only in and through the help of a cooperative community (Savage 2002: 93). What is distinctive of democratic life is, therefore, the development of a "habit of amicable cooperation" through which citizens cooperate and justify their preferences concerning the public interest and not on individualistic terms (Dewey, 1981a: 227).

While the relational view of democracy is often presented in opposition to liberal democracy, Dewey did not reject liberal values. In his view, the realization of individual autonomy and freedom presupposes the existence of a collective community

---

[4] Relational accounts of democracy have been presented by Elizabeth Anderson (1999, 2009), Samuel Scheffler (2010,) and Niko Kolodny (2014)

in which individuals can flourish. From Dewey's standpoint, therefore, the debate between liberals and communitarians that occupied Western political philosophers for most of the 1980s and the 1990s rests on a false dichotomy. In his view, "[t]he real problem comes from supposing that we must choose between individual autonomy and genuine community" (Savage, 2002: 93). Individual liberties and community life are intrinsically intertwined such that the cultivation of one presupposes the cultivation of the other.[5]

Thus relational democrats reject the traditional liberal view of democracy. What is distinctive of traditional views of liberal democracy is the belief that liberal principles must shape and limit democratic rule. Most liberals maintain that the 'government of the people, by the people, for the people' should be directly promoted. Democracy recognizes the self-government authority of individuals by allowing individuals to (more or less directly) govern themselves. Nevertheless, the liberals' support for democracy is not unconditional: to be "conducive to freedom", in Saffron and Urbinati's words, democratic rule must be shaped and defined by liberal values (2013: 443). Relational democrats do not deny the value of liberal principles but question their role in a liberal democratic society. Such principles should be pursued for the sake of the people, such that the people, not the realization of certain states of affairs, are the ultimate objective of interest for democracy (Anderson, 2009: 223). The importance of a cooperative community for the development of individuals redefines the meaning and the goals of democracy. The protection of individual liberties, although valuable for a democratic community, is not the ultimate justification of a democratic society.

Through the lens of relational democracy, democratic politics acquires a new meaning. Dewey maintains that democracy must adopt a scientific attitude which he calls the 'experimental approach' (Dewey, 1981b: 167). Democratic politics is the process through which the community identify what issues are collective problems and puts together different epistemic resources to solve them. Democratic politics is, therefore, a form of 'social inquiry'. Its decision making is a constant and neverending process, expanding beyond the short-term electoral cycles. Even if a political decision proves to be successful, it could lead to new problems and perhaps the need to consider different perspectives. The goal-oriented and epistemic value of democratic-decision making distinguishes the relational understanding of democratic politics from classic liberal accounts. The latter stress the justifiability of a political decision, while the relational democrats view democracy as first and for most a process where intelligent decisions are made. In democracy, experimenting with new solutions and mistakes are remedied by pulling together different epistemic inputs from members of society (Anderson, 2006).

---

[5] This idea is well captured in Dewey's words: "[l]iberty is that secure release and fulfillment of personal potentialities which take place only in rich and manifold association with others: the power to be an individualized self making a distinctive contribution and enjoying in its own way the fruits of association" (Dewey, 1946: 150).

3.3   Relational-democratic Approach to New Digital Technologies

The relational understanding of democratic politics redefines the relationship between democracy and technological experimentations, such as those we have recently undergone. From a relational-democratic perspective, in times of crisis, experimentation with new technologies is a possibility open for discussion. Depending on the nature of the challenge that the community faces, even the introduction of digital surveillance technologies can become a topic of public discussion provided that they can help the community to fulfil its collective aim. However, there is a caveat: the relational ideal of democracy does not support the introduction of digital surveillance under all conditions. For pragmatists, democracy is ultimately a community of equals and, therefore, even democratic experimentations with new technologies that may well have a significant impact on society must not be insulated from public deliberation. Furthermore, for pragmatists, an undemocratic decision would also have fewer chances to reach the 'best' decision. Reaching a decision independently from the public discussion can insulate the decision-making from key contextual information and the epistemic inputs of different members of society that may hold the key to finding a successful solution in a given situation.[6]

Having clarified the relational democracy's approach to experimentation with new technologies, it remains unclear how society can initiate such experiments and avoid the potential ethical and practical risks of technological experimentations. Assuming that a democratic society can approve the introduction of digital surveillance technologies, do democratic societies have the means and know-how to manage these technologies? Is it realistic to believe that such experiments would not be abused by self-interested parties? As mentioned above, this hypothesis is not far-fetched if we consider real examples of commercial digital surveillance, like the one practised by Facebook and Google.

One way to approach this question is with recourse to Ibo van de Poel's ethical framework for evaluating experimental technology. Van de Poel argues that we have limited operational experience of certain technologies, such that their social benefit or threat cannot be straightforwardly being accessed (2016). Anticipatory methods to predict the social impact of experimental technologies are likely to be only partially successful. The outcome of the introduction of an experimental technology depends on how the technology 'connects' with a given social context, however, our ability to foresee the effect of such connection is limited since we have minor

---

[6] Contemporary epistemic democratic theories and pragmatism converge on the claim that democracy's value partly depends on its ability to reach 'good' decisions. However, they hold different views of what a 'good' political decision is. While Dewey maintains that the epistemic power of democracy depends of its ability to meet our own reflective satisfaction with the practical results, epistemic democrats maintain that democracy can "track" or "correspond" to truth. For a defense of Dewey's idea of successful decision making against epistemic democratic approaches, see Fuerstein (2021), while for an overview of the epistemic democratic debates on the true-tracking property of democracy, see Landemore (2017).

operational experience of this technology. Thus, surprise and uncertainty about the introduction of these technologies will remain. Furthermore, "anticipation may well lead to a focus on scenarios that are morally thrilling but very unlikely" (van de Poel, 2016: 670). This also suggests that any adoption of these technologies is "de facto experimentation" (van de Poel, 2016: 672) and that a different method to appraise new technological developments from anticipatory studies is in order.

Van de Poel's analysis of experimental technologies is valuable for this chapter because digital surveillance technologies can be considered as a type of experimental technology. The difference between experimental and non-experimental technology primarily depends on the operational experience that we have of that particular technology. Furthermore, "how much and for how long a period, operational experience is required may well depend on the technology and the kind of (social) impacts one is interested in or worried about" (van de Poel, 2016: 670). Arguably surveillance is not a new phenomenon, but digital surveillance is. Our operational experience of this new form of surveillance remains quite limited, especially in emergency situations like a health crisis.

So, assuming that a democratic society can democratically approve the introduction of digital surveillance technologies, how can such a society control the introduction of these technologies? To address the issue of controlling experimental technologies, Van de Poel suggests monitoring the social effects of the new technologies when they are gradually introduced into society and improvements to the technology can be made accordingly (van de Poel, 2016: 670). Drawing on Dewey's approach to social experimentation, van de Poel proposes a set of ethical general principles to guide the introduction of new technologies: non-maleficent, beneficence, respect for autonomy, and justice. The non-maleficence principle requires the prevention of harm in so far as it is reasonably possible and to stop or reduce the damages if harm occurs (van de Poel, 2016: 678). Indeed, it would be unreasonable to require that no harm is caused because social experiments with new technology could give rise to unknown harm (van de Poel, 2016: 678). While the beneficence principle requires new technologies to add value to society, the principles of justice and respect for autonomy entail that social experimentations must be carried out while respecting the procedural justice and the autonomous choice of a group (van de Poel, 2016: 676–77).[7]

Critics may welcome van de Poel's ethical framework but complain that the relational idea of democratic experimentation with new technologies remains quite abstract. Assuming that democracy is 'a way of life' and the ethical principles suggested by van de Poel can be adopted to experimentally introduce and monitor new technologies, how would this democratic experimentation process look like in practice? For my relational democratic argument to work, I will explain how relational democratic experimentations with technologies would look like in such an

---

[7] In the pragmatic spirit, van de Poel stresses that these principles are not set in stone. They remain open to specification and possible revision according to the specific context of implementation (van de Poel, 2016: 684).

emergency. In the next section, I will discuss the case of South Korea's experimentation with new digital surveillance technologies during the first year of the pandemic.

3.4 The Case of South Korea

Several reasons suggest that South Korea is a good case study for the purpose of this chapter. Firstly, South Korea's population (~51 million) is similar to the Western democracies of medium size, such as Italy (~60 million population), Spain (~47 million population), and England (~56 million population). Secondly, a democratic political and legal framework shape the democratic life of South Korea. Thirdly, the relationship between the relational model of democracy and real forms of democracy in East Asia has been debated for a long time by East-Asian democratic theorists. Among all forms of democracy, several East-Asian scholars consider the relational conception of democracy to be the most compatible with the Confucian values and ideals that continue to shape the socio-political lives of contemporary East Asia.[8]

Despite being one of the first countries to experience a COVID-19 outbreak, South Korea is one of the countries that dealt with the pandemic most swiftly and efficiently. The South Korean containing strategy was defined as "a success" by international media and the term 'K-quarantine' has become synonymous with the South Korean successful management model (Yang, 2021). The results of the South Korean approach to the pandemic are impressive especially if they are compared with those of Western liberal democracies of similar sizes, such as Italy or Spain. These two countries detected their first COVID-19 patient almost one month after the South Korean's first COVID-19 patient and they went on to see far more deaths and cases than South Korea. On 1st January 2021, South Korea reported a total of ~62.000 cases and ~900 deaths (Worldometers, 2021a), while Italy counted more than 2 million cases and almost 75.000 deaths. Spain logged almost 2 million cases and around 60.000 deaths (Worldometers, 2021b, 2021c). In terms of social restrictions and limitation of movements, South Korea only enforced a partial lockdown and did not close its economy nor its borders. On the contrary, Italy experienced one national lockdown that lasted more than two months while Spain went through a lockdown that lasted 3 months. In 2020, the Korean GDP contracted by 2%, while the GDP of Italy and Spain contracted by almost 10% (European Commission (2021a, 2021b).

Several experts attribute South Korea's success in managing the COVID-19 pandemic to three main factors: (a) learning from the history of respiratory diseases, (b) an experimental approach to technologies within the limits imposed by a

---

[8] For instance, both relational democracy and Confucianism assume a relational conception of the self and value the relationship between citizen and state as valuable for non-instrumental reasons. For a detailed analysis of this issue, see Tan (2003).

democratic legal framework, and (c) social ethos. These considerations make South Korea's experiments with new technologies relevant to the discussion on the experimental relational democratic approach to techno-politics that could be adopted in emergencies.

3.4.1 Learning from the History of Respiratory Diseases

In 2015, South Korea was jolted by the MERS (the Middle East Respiratory Syndrome), which resulted in 36 deaths. This number may appear small to readers that have lived through the COVID-19 pandemic, but at that time it led to a strong public outrage in South Korea. The outbreak cost an estimated loss of US$2.6 billion in tourism revenue and almost US$1 billion on diagnosis, treatment, and other parts of its response. The mismanagement of the MERS outbreak was also one of the reasons for the election loss of the incumbent government (Oh et al., 2020). During the MERS emergency, the government shared information only among expert groups while keeping the public in the dark on several aspects of the crisis management (e.g. civilians were not aware of which hospitals were treating MERS patients). This secrecy made the handling of the emergency difficult for the government which soon lost control of the situation.

South Korea drew on the lessons learnt from this tragic experience in managing the COVID-19 pandemic. Right after the MERS crisis, the new South Korean government proposed 48 reforms to improve public health emergencies in the control of diseases and response to a pandemic. These reforms included the possibility for the government to collaborate with the private sector in the deployment of new digital surveillance technologies for health emergencies. I will return to the specification of these new technologies shortly. For now, I want to point out that, because these reforms were democratically turned into law, their introduction respected two of the general principles listed by van de Poel: justice (procedural justice) and autonomy (the autonomous choice of a group). Furthermore, their introduction mirrors the democratic idea that we discussed in the previous section. The decision to adopt digital surveillance technologies during an epidemic was the outcome of a democratic debate.

The new government also learned its lesson on public communication. Unlike the secretive approach adopted during the MERS epidemic, total transparency became the key for the government's communication with the public in the COVID-19 pandemic. As I explain later, this strategy had some negative consequences, but from a general perspective, the South Korean gradual experimentation with new technologies during the COVID-19 crisis is in line with the relational idea of democratic decision-making which aims to learn about what works and, at the same time, define the conditions under which a solution can be seen as working from the citizens' perspective (Anderson, 2009: 217).

3.4.2 Experimentation with New Digital Surveillance Technologies

As discussed in the previous section, some of the laws that the South Korean national assembly passed after the MERS emergency concern the regulation on the use of digital surveillance technologies in emergencies. These new digital surveillance technologies infringe the privacy and freedom of citizens because they gave access to the government to the private information of the citizens although the government committed not to reveal this information to the public.

Distinctive of the South Korean case is the number of digital surveillance technologies that were deployed at the same time and their areas of coverage. To control the spread of the COVID-19 virus, the South Korean government-commissioned private businesses to develop applications and online tracing maps to monitor the movements of COVID-19 patients who were supposed to be in self-isolation, to identify the persons who had come into contact with COVID-19 patients and share information on the crisis management (such as the supply of masks). These platforms gathered data through four main types of surveillance technological strategies: control of phone geolocation data, credit card location data, transportation card data, and CCVT footage. Through these technologies, health authorities could contact and trace thousands of potential patients, and test and isolate patients before they could unknowingly infect others.

To facilitate the identification of potential cases during the early stages, the Infectious Disease Control and Prevention Act was revised after the MERS crisis. The new document allows the government to collect citizens' data, while at the same time it guarantees South Koreans the right to be informed on what data the government is collecting about them. "This Act, therefore, serves as a social contract between the state and Korean citizens to control the use of tracking technologies" (Schwak, 2020: 19). However, in some cases, the case-related information that the government shared with the citizens was sufficient for some members of the public to determine the patients' identities as the information that was shared initially with the public included the patients' ages, the blocks of apartments where they lived, the names of the places they had visited recently, details on how they became infected, and where they were tested and treated (Yang, 2021). This allowed individuals who visited the same places to be quickly informed and tested, but it also contributed to the rise of malicious comments online on what the public perceived as irresponsible choices, like patients' decisions to visit many public places in one day. As Korea's number of deaths for COVID-19 remained low, many people became more afraid of online criticism than contracting the virus (BBC, 2020).

This online social stigmatization highlights the negative effect of the South Korean experiment with new digital surveillance technologies. Although it did not escalate into physical harm, it reportedly caused psychological harm to many COVID-19 patients who suffered cyberbullying. This phenomenon, therefore, reveals an unintentional breach of the privacy of the experimental subjects – a specification of van de Poel's principle of non-maleficence (2016: 679). However, the response of the

South Korean health authorities to social stigmatization and cyberbullying also goes some way to fulfiling van de Poel's principle of non-maleficence. This non-maleficence principle requires preventing harm as far as possible and suspending the experiment or taking measures to reduce harm. The National Human Rights Commission of South Korea took steps to address the rise of a digital 'witch hunt' and requested the government to revise its data management policy to ensure anonymity and protect the mental health of the COVID-19 patients (Schwak, 2020: 20). The Center for Disease Control and Prevention followed suit and issued new guidelines for patient data collection and disclosure: it decided to exclude personally identifiable information (such as work and home addresses) from public discourse, limit the patients' logs from one day before the symptoms occurred to the date of quarantine (or if asymptomatic, one day before the quarantine), and determine the range of contacts traced based on the patient's symptoms, exposure conditions, and timing (Jo, 2020). These decisions were based on the joint effort of the Korean state and other stakeholders to act according to the non-maleficence principle and rectify the damages caused by the introduction of the new technologies.[9]

The South Korean case illustrates how democratic experimentation with new digital surveillance technologies was carried out. As we have learned, such experimentation was not always smooth and despite its material benefits it was indeed also characterized by unforeseen negative effects and detrimental social phenomena. However, it is also an example of progressive and democratic decision-making process that learned from its mistakes and addressed new problems in ways it saw fit. More importantly, this progressive decision-making process did not follow a top-down approach; the public was indirectly involved in the process through democratic representation and more directly in providing epistemic input in the digital public sphere. This suggests not only technological experimentation was being monitored but also there was effective communication between the government and the members of the public.

My analysis of the South Korean case, based on Van de Poel's autonomy principle, reveals one shortcoming of the Korean digital surveillance experiment: the experimental subjects were not able to withdraw from the experiment (condition 13 of Respect for Autonomy, Van de Poel 2016: 680). However, the patients could submit a petition to review their logs. Unlike many Western liberal democracies, there was no public outrage in South Korea over the government's deployment of new digital surveillance technologies. According to a survey carried out in June 2020, South Koreans' valuation of their government responses to COVID-19 was

---

[9] Another important aspect of these reforms that South Korea approved after the MERS crisis concerned the reorganisation of the South Korean National Infection Prevention and Control System for the Purpose of Immediate Response to Emerging Infectious Diseases. These reforms ranged from the increase of the number of initial response systems to respond to an outbreak of emerging infectious diseases, to the establishment of a 24-h-a-day Emergency Operations Center to collect and monitor information on infectious diseases in real-time, a specialized diagnosis and treatment system with quarantine and isolation facilities to detect and prevent the outbreak of emerging infectious diseases, and the strengthening of the interactive telemedicine system (South Korean Ministry of Health and Welfare, 2015).

the highest in the world (74%) after the one of mainland Chinese people (80%) (Lazarus et al., 2020). There may well be many reasons for this. First, in South Korea, the use of these technologies was regulated by laws that were democratically approved by the representative chambers and this, in turn, contributed to the public trust in the government's management of these technologies. Second, because the experience of the MERS crisis was still vivid in the memory of many South Koreans, many citizens deemed the temporary curb on their liberties as a necessary evil to control the pandemic. Third, experts believe that South Koreans' acceptance of digital surveillance technologies during the health crisis may be due to their social ethos, a distinctive aspect of their public culture.

3.4.3 Social Ethos

Communal values in South Korea were a big part of its success in the management of the pandemic. The introduction of the new technologies took place in a societal context that was characterized by strong public communal values (Stockwin, 2020). A high level of civic solidarity is suggested by the fact that 93% of the South Korean citizens maintained that they were practising social distancing well (Jaung, 2020). Scholars believe that social cohesion is a common characteristic in most of the East Asian region. According to Yves Tiberghien, "[i]n all East Asian countries, saving lives during a natural disaster is seen as the primary duty of the government" and the roots of these trends go back to historical and cultural factors and, perhaps, the long influence that Confucianism had in the region (Tiberghien, 2021: 31). Despite its ethnic and cultural diversity, there is a shared belief among East Asian societies that, "[w]hen a crisis hits, society must pull together" (Tiberghien, 2021: 37). This general belief, together with the early mobilization of centralized pandemic command centres, and the very high and general adoption of masks, allowed several East Asian countries to perform better than what observers expected (Tiberghien, 2021: 44).

Furthermore, the importance of the context in which the new technology was successfully used is evident in the South Korean case. The same success with technological experimentation would not have been possible without South Korea's digital infrastructure. At the beginning of the pandemic, South Korea was a highly technological country; it has the world most extensive broadband and mobile network. Almost all South Korean citizens own mobile phones, with 95% owning smartphones. Approximately 860,000 4G and 5G transceivers, which cover the entire country, record phone locations automatically with complete accuracy. In addition, in 2015, almost 1.5 million CCTVs covered public and private places (Yang, 2021).

Besides digital development, a second key aspect of the South Korean success concerned the democratic aspect of such an experiment. As we said before, social trust was reinforced by the democratic procedure through which such experimentation was legally approved. As pointed out by Juliette Schwak: "[i]t is Korea's democracy that has proved efficient, rather than technology per se. If lessons must be drawn, foreign observers should be wary of picking tracking technologies as the only solution to the current health crisis" (Schwak, 2020: 21). The case of South Korea is ultimately a case of state digital surveillance, but the discussion of this case does not aim to defend forms of state digital surveillance. On the contrary, the aim is to explain how a democratic society can manage experimentations with digital surveillance by bringing the latter under democratic control.

3.5 Conclusive Reflection

How should democratic societies experiment and control digital surveillance technologies? This question has become more pressing than ever with the COVID pandemic, where different states around the world have implemented different approaches to digital surveillance in their battle against COVID. This chapter has argued that self-government is the core principle of democratic government, thus democratic societies must bring digital surveillance under the control of democratic institutions and the relational ideal of democracy is a useful paradigm from which a democratic approach to digital surveillance democracy can be developed. The relational ideal suggests a context-sensitive approach to experimentation, in which input from members of the public and public deliberations are key to managing technologies. To clarify my claim, I have discussed the case of South Korea's experimentation with new digital surveillance technologies to explain how this can be realized. The relational understanding of democracy does not deny the value of individuality. It aims to complement the liberal understanding of democracy, not compete with it. So going forward, we should not reject liberal values, but we need to re-assess their meanings. A change in the way we conceptualize democracy can not only mark a theoretical turning but also examine how democracy is practised. In other words, it calls us to revise our approach to politics as citizens and to transform the way we 'do' democratic politics.

More research needs to be done to define the exact value of the relational democratic model for experimentation with digital surveillance technologies. We should clarify what other approaches to digital surveillance technologies can be derived from alternative conceptions of democracy and then compare their strengths to those of relational democracy.

References

Anderson, E. (1999). What is the point of equality? *Ethics, 103*(9), 287–337.
Anderson, E. (2006). The epistemology of democracy. *Episteme, 3*(1–2), 8–22.
Anderson, E. (2009). Democracy: Instrumental vs non-instrumental value. In T. Christiano & J. Christman (Eds.), *Contemporary debates in political philosophy*. Wiley-Blackwell.

BBC. (2020). Coronavirus privacy: Are South Korea's alerts too revealing? *BBC News*. Available at: https://www.bbc.com/news/world-asia-51733145. Accessed on 09 March 2021.

Bogard, W. (1996). *The simulation of surveillance: hypercontrol in telematic societies*. Cambridge University Press.

De La Garza. (2020). Contact tracing apps were big tech's best idea for fighting COVID-19. Why haven't they helped? *Times*. Available at: https://time.com/5905772/covid-19-contact-tracing-apps/. Accessed on 25 November 2021.

Dewey, J. (1938/1991). Democracy and education in the world today. In J. A. Boydston (Ed.), *John Dewey: The later works* (Vol. 13). Southern Illinois University Press.

Dewey, J. (1946). *The public and its problems: An essay in political inquiry*. Gateway Books.

Dewey, J. (1981a). Creative democracy: The task before us. In J. A. Boydston (Ed.), *The later works of John Dewey, 1925–1953* (Vol. 14, pp. 224–230). Southern Illinois University Press.

Dewey, J. (1981b). Freedom and culture. In the later works of John Dewey, 1925–1953, vol. 13. In J. A. Boydston (Ed.), *The later works of John Dewey, 1925–1953* (Vol. 14, pp. 65–188). Southern Illinois University Press.

European Commission. (2021a). *Economic forecast for Italy*. Available at: https://ec.europa.eu/info/business-economy-euro/economic-performance-and-forecasts/economic-performance-country/italy/economic-forecast-italy_en. Accessed on 01 September 2021.

European Commission. (2021b). *Economic forecast for Spain*. https://ec.europa.eu/info/business-economy-euro/economic-performance-and-forecasts/economic-performance-country/spain/economic-forecast-spain_en. Accessed on 01 September 2021.

Fuerstein, M. (2021). Epistemic democracy without truth: The Deweyan approach. *Raisons Politiques, 81*(1), 81–96.

Gilliom, J. (2001). *Overseers of the poor: Surveillance, resistance, and the limits of privacy*. University of Chicago Press.

Graham, S., & Wood, D. (2003). Digitizing surveillance: Categorization, space, inequality. *Critical Social Policy, 23*(2), 227–248. https://doi.org/10.1177/0261018303023002006

Jaung, H. (2020). Another Korean miracle? The paradox of civic trust in the Dataveillance state in the COVID 19 crisis. *Edinburgh Forum on Korea*. Available at: https://blogs.ed.ac.uk/edinburghforumonkorea/2021/05/31/another-korean-miracle-the-paradox-of-civic-trust-in-the-dataveillance-state-in-the-COVID-19-crisis/. Accessed on 30 March 2021.

Jo, E. A. (2020). South Korea's experiment in pandemic surveillance (April 13, 2020). *The Diplomat*. Available at: https://thediplomat.com/2020/04/south-koreas-experiment-in-pandemic-surveillance/. Accessed on 01 October 2021.

Kolodny, N. (2014). Rule over none II: Social equality and the justification of democracy. *Philosophy & Public Affairs, 42*(4), 287–336.

Landemore, H. (2017). Beyond the fact of disagreement? The epistemic turn in deliberative democracy. *Social Epistemology: A Journal of Knowledge, Culture and Policy, 31*(3), 277–295. https://doi.org/10.1080/02691728.2017.1317868

Lazarus, J., Ratzan, S., Palayew, A., Francesco, B., Binagwaho, A., Spencer, K., et al. (2020). COVID-SCORE: A global survey to assess public perceptions of government responses to COVID-19(COVID-SCORE-10). *PLoS One, 15*(10), e0240011. https://doi.org/10.1371/journal.pone.0240011

Lewis, D. (2021). Contact-tracing apps help reduce COVID infections, data suggest. *Nature*. Available at: https://www.nature.com/articles/d41586-021-00451-y. Accessed on 25 November 2021.

North Dakota. (2020). North Dakota announces launch of Care19 Alert app to help reduce spread of COVID-19 as students return. *North Dakota Health*. Available at: https://www.health.nd.gov/news/north-dakota-announces-launch-care19-alert-app-help-reduce-spread-covid-19-students-return. Assessed on 25 November 2021.

Oh, J., et al. (2020). National response to COVID-19 in the Republic of Korea and lessons learned for other countries. *Health Systems and Reform, 6*(1), e1753464. https://doi.org/10.1080/23288604.2020.1753464

Saffon, M. P., & Urbinati, N. (2013). Procedural democracy, the bulwark of equal liberty. *Political Theory, 41*(3), 441–481.

Savage, D. (2002). *John Dewey's liberalism individual, community, and self-development.* Southern Illinois University..

Scheffler, S. (2010). *Equality and tradition: Questions of value in moral and political theory.* Oxford University Press.

Schwak, J. (2020). A democratic *tour de force*: How the Korean State successfully limited the spread of COVID-19. *Asie. Visions* 117, Ifri.

Simons, J., & Ghosh, D. (2020). *Utilities of democracy: Why and how the algorithmic infrastructure of facebook and google must be regulated.* Brookings Institute.

South Korean Ministry of Health and Welfare. (2015). Measures to reform national infection prevention and control system for the purpose of immediate response to emerging infectious disease. *Press release.* Available at: https://www.mohw.go.kr/eng/nw/nw0101vw.jsp?PAR_MENU_ID=1007andMENU_ID=100701andpage=1andCONT_SEQ=326060. Accessed on 01 September 2021.

State of South Dakota. (2020). *Care19 app, COVID-19 in South Dakota.* Available at: https://covid.sd.gov/care19app.aspx. Assessed on 25 November 2021.

Stockwin, A. (2020). How will COVID-19 reshape the world? *East Asia Forum.* Available at: www.eastasiaforum.org/2020/04/15/how-will-COVID-19-reshape-the-world/. Accessed on 01 October 2021.

Tan, S.-H. (2003). *Confucian democracy: A Deweyan reconstruction.* SUNY Press.

Tiberghien, Y. (2021). *The east Asian COVID-19 paradox.* Cambridge University Press.

van de Poel, I. (2016). An ethical framework for evaluating experimental technology. *Science and Engineering Ethics, 22,* 667–686. https://doi.org/10.1007/s11948-015-9724-3

Worldometers. (2021a). COVID-19 conoronavirus pandemic, South Korea. *Worldometer.info.* Available at: https://www.worldometers.info/coronavirus/country/south-korea/. Accessed on 27 August 2021.

Worldometers. (2021b). COVID-19 conoronavirus pandemic, Italy. *Worldometer.info.* https://www.worldometers.info/coronavirus/country/italy/. Accessed on 27 August 2021.

Worldometers. (2021c). COVID-19 conoronavirus pandemic, Spain. *Worldometer.info.* Available at: https://www.worldometers.info/coronavirus/country/spain/. Accessed on 27 August 2021.

Yang, M. (2021). Behind South Korea's success in containing COVID-19: Surveillance technology infrastructures. *Items: Insights from the Social Sciences.* Available at: https://items.ssrc.org/COVID-19-and-the-social-sciences/COVID-19-in-east-asia/behind-south-koreas-success-in-containing-COVID-19-surveillance-technology-infrastructures/. Accessed on 01 October 2021.

Zuboff, S. (2019a). *The age of surveillance capitalism: The fight for a human future at the new frontier of power.* Public Affairs.

Zuboff, S. (2019b). Surveillance capitalism and the challenge of collective action. *New Labor Forum, 28*(1), 10–29.

Open Access   This chapter is licensed under the terms of the Creative Commons Attribution 4.0 International License (http://creativecommons.org/licenses/by/4.0/), which permits use, sharing, adaptation, distribution and reproduction in any medium or format, as long as you give appropriate credit to the original author(s) and the source, provide a link to the Creative Commons license and indicate if changes were made.

The images or other third party material in this chapter are included in the chapter's Creative Commons license, unless indicated otherwise in a credit line to the material. If material is not included in the chapter's Creative Commons license and your intended use is not permitted by statutory regulation or exceeds the permitted use, you will need to obtain permission directly from the copyright holder.

Chapter 4
Contact Tracing Apps for the COVID-19 Pandemic: A Responsible Innovation Perspective

George Ogoh, Simisola Akintoye, Damian Okaibedi Eke, Tonii Leach, Paschal Ochang, Adebowale Owoseni, Oluyinka Oyeniji, and Bernd Carsten Stahl

4.1 Introduction

The COVID-19 pandemic has precipitated the first real opportunity to test the efficacy of the Responsible Research and Innovation framework (RRI) in a global health crisis. Although the European Commission has promoted RRI since 2011, little is known about the application of RRI approaches in a health crisis. This is especially important as high levels of both infection and death, along with the difficulty in finding a completely successful treatment for COVID-19, has paved the way for bold new approaches to health research and innovation. One such approach which has received a lot of attention during the COVID-19 health crisis is digital contact tracing applications (CTA). This chapter provides an extensive assessment of RRI related issues during the development of CTAs, discussing these issues from the experience of four countries – Germany, France, Spain, and the United Kingdom (UK) – and shows that although they did not explicitly use the RRI approach during the development of their CTAs, some of their activities during this period can be mapped to RRI. We ask: 'What elements of RRI are identifiable in the development of contact tracing apps during the COVID-19 health crisis?

Although contact tracing is a well-established evidence-based public health measure for responding to outbreaks of infectious disease (Riley et al., 2003; World Health Organisation WHO, 2014; Kwok et al., 2019), digital CTAs were first developed in response to COVID-19. They have since raised serious ethical concerns. For example, in response to the planned release of a CTA in the UK, the Nuffield

G. Ogoh (✉) · S. Akintoye · D. O. Eke · T. Leach ·
P. Ochang · A. Owoseni · O. Oyeniji · B. C. Stahl
De Montfort University, Leicester, UK
e-mail: george.ogoh@dmu.ac.uk; simi.akintoye@dmu.ac.uk; damian.eke@dmu.ac.uk;
antonia.leach@dmu.ac.uk; P2614609@my365.dmu.ac.uk; adebowale.owoseni@dmu.ac.uk;
P2623648@my365.dmu.ac.uk; bstahl@dmu.ac.uk

© The Author(s) 2022
M. J. Dennis et al. (eds.), *Values for a Post-Pandemic Future*, Philosophy
of Engineering and Technology 40, https://doi.org/10.1007/978-3-031-08424-9_4

Council on Bioethics (2020) raised twenty questions about this application, including questions on privacy, security and ethics. Similarly, academics at Oxford University have provided 16 questions for the ethical assessment of CTAs (Morley et al., 2020), and tech experts in the United States (U.S) have suggested that this technology raises questions of reliability and inaccuracy of information (Sterman & Brauer, 2020). RRI, we contend, can help to enable a better understanding of the types of concerns highlighted here, and opportunities to mitigate them. Intending to mitigate societal concerns of emerging technologies, the European Commission began promoting RRI to enable a better understanding of unintended impacts of innovation whilst minimising associated ethical issues. RRI suggests that this can be achieved by bringing greater democracy to science and technology through research and innovation processes that emphasise public participation, deliberation, and reflexivity (Von Schomberg, 2011).

This chapter, therefore, highlights the issues around the development of COVID-19 CTAs and draws attention to salient issues regarding RRI in crises. The issues encountered during the development of contact tracing apps by four governments are described and an indication of the implications for the application of RRI in the development of ICTs during health crises is provided.

4.2   RRI for Crisis Response and Management

Emerging technologies are unpredictable. It is challenging to fully understand the ramifications of their adoption, trajectories, and societal acceptability. Unsurprisingly, a 'policy vacuum' (Moor, 1985; Moor, 2005) is commonly present in the governance of emerging technologies. Policies are often crafted in an institutional void without adopting generally accepted rules and norms (Hajer, 2003). To combat such issues in the European science and innovation arena, an approach of RRI was formally proposed by the European Commission (EC) in 2011 and subsequently adopted (European Commission, 2011).

RRI has been described as a 'transparent, interactive process by which societal actors and innovators become mutually responsive to each other with a view on the ethical acceptability, sustainability, and societal desirability of the innovation process and its marketable products' (Von Schomberg, 2011). To this end, the EC has promoted several fundamental elements as critical actions for RRI, including public engagement, gender equality, science education, open access, ethics, and governance (European Commission, 2015). Thus, RRI may be characterised by its focus on ethical aspects, meeting societal expectations, and inclusive participation.

As early as 2013, shortly after the formal ratification of the RRI framework, its usefulness as an approach for identifying the profound impacts of technology during crises was recognised. Stilgoe et al. (2013) argued that the 2008 financial crisis, (the most contemporaneous example of a crisis with wide-reaching implications) was an example of disruptive situations where RRI could have made a significant difference. They suggested this because existing governance processes, often premised on formal risk assessment, have done little to identify many of the profound impacts of innovation that have plagued society.

Surprisingly, however, since then, little has been said about using the RRI framework in crises, and only a handful of authors have highlighted its applicability in crisis management. One example is Buscher et al. (2018), who highlight the usefulness of RRI for crisis and disaster risk management, arguing that disaster risk management models are changing from publicly-funded command and control to 'datafied' and netcentric approaches with increased monitoring and surveillance, raising profiling and social sorting concerns. They suggest that the application of RRI to the development of information technologies (IT) for crisis and disaster management can help maximise the potential benefit of IT, address social concerns, and ensure social value alignment. To enable this process, they started an initiative that primarily has brought together 'a critical mass of stakeholders' for co-creation of principles, knowledge exchange, critical dialogue around controversies and standards for responsible IT research and innovation in disaster risk management.

It has also been suggested that the RRI framework has practical implications for the public health crisis triggered by the Syrian War. Khallouf (2018), who made the call for the urgent application of RRI in this context, suggests that the interdisciplinary collaborative approach which RRI promotes could help ensure that cloud computing systems developed for improving health care delivery are sustainable, ethically acceptable, and socially desirable. During the COVID-19 health crisis, the applicability of RRI has also been recognised. Braun et al. (2020) opened the dialogue on Responsible online Research and Innovation (RoRI) to deliberate on the challenges and socio-ethical opportunities that the use of online tools in place of face-to-face interactions has brought. They maintain that it is vital to consider an RRI perspective on the 'onlineification of everything' as it is easy for research to get hijacked by corporate interests leading to an obstruction of inclusive and democratising dynamics. To do this, they suggest that the procedural heuristic proposed by Stilgoe et al. (2013) based on the four dimensions of anticipation, inclusion, reflection, and responsiveness should be used alongside the RRI keys proposed by the EC.

One application of the Responsible Research and Innovation framework during the COVID-19 crisis can be seen in the Human Brain Project (HBP), where it has been foundational in promoting digital inclusiveness when people are required to work from home. Grasenick and Guerrero (2020), who introduced this concept, started 'i-Include', an initiative for inclusive digital engagement developed to ensure that no one is left behind when increasing the virtualisation of work, meetings, and association and that issues around diversity are also considered in digital collaborations. To this end, they introduced a set of recommendations for social and family life, stress and anxiety, roles and responsibilities in different career stages, as well as team cohesion and virtual collaboration.

In the context of this discourse, however, one of the most relevant applications of RRI in crises is highlighted by Monteiro et al. (2017), who considered the response to the Zika virus outbreak in Brazil. They maintain that in attempting to respond quickly to emergent health crises, irresponsibility could arise in implementing science and technology. Irresponsibility, they argue, comprises "forms of crisis governance implemented in times of emergency which do not fully engage with the public in ways which may be considered participatory or reflexive, a lack of care for the future, and a lack of reflexiveness about said solutions." They argue for a balance between vigilance in times of crisis and responsible research and innovation in everyday situations. They highlighted how debates for the adoption of controversial technologies in the health crisis failed to consider pre-existing unequal social relationships and broader socio-political issues.

Nevertheless, their discussion highlighted the failure of crisis governance to engage with the public in participatory and reflexive ways during the development of solutions for the health crisis. This highlights that the 'transparent, interactive process by which societal actors and innovators become mutually responsive to each other' for which Von Schomberg (2011) alluded to in his definition of RRI was not applied, resulting in controversial solutions being promoted.

In summary, this analysis shows that RRI for crisis management appears to be characterised by calls or initiatives for wider and more inclusive participation in the management of crisis. The following examples highlight this:

1. Buscher et al. (2018) started an initiative involving a 'critical mass of stakeholders' for co-creation and critical dialogue to highlight the usefulness of RRI for crisis and disaster risk management.
2. Khallouf (2018) called for interdisciplinary collaboration to develop cloud computing systems in response to the health crisis triggered by the Syrian War.
3. Monteiro et al. (2017) highlighted the failure of crisis governance to engage with the public in participatory and reflexive ways and called for more to be done in this area.
4. An initiative for inclusive digital engagement was started in a large interdisciplinary project to help address issues of participation for those working from home during the COVID-19 pandemic (Grasenick & Guerrero, 2020); and
5. Braun et al. (2020) highlighted how corporate interests could lead to an obstruction of inclusive and democratising dynamics.

Inclusive participation features prominently in the definition of RRI, dimensions of RRI, and the RRI keys proposed by the EC. This indicates the importance of highly inclusive processes for responsible innovation. This chapter expands on Monteiro et al.'s (2017) approach by assessing how anticipation, inclusive participation and reflexivity may have affected the development of CTAs.

4.3   Contact Tracing and the Move to CTAs

Contact tracing, alongside testing and vaccination, is a critical approach for infectious disease case management. It is used to identify, isolate and provide support to individuals who have been in contact with people with infectious diseases of concern such as smallpox, tuberculosis, Ebola (Crook et al., 2017) and STDs (Hogben et al., 2016). The logic behind contact tracing is that when a person tests positive for infectious disease, possible contacts are identified, notified and advised on any additional medical interventions. Conventionally, this was done via interviewing the index case followed by telephone calls or visits to the identified contacts. For several reasons, digital contact tracing has been heralded as pivotal in the fight against COVID-19. First, COVID-19 has a very high infection rate and has "tricky and complex mechanisms that have facilitated its rapid and catastrophic spread worldwide" (Pitlik, 2020). This makes it necessary to adopt faster means of breaking the chain of transmission. Second, the availability of technologies such as mobile and internet services, AI, Machine learning and other data-driven tools can help healthcare systems to achieve faster contact tracing to match the rate of infection (van der Schaar et al., 2021; Cave et al., 2021). The deployment of these tools at speed and scale for contact tracing has significantly accelerated since the global spread of COVID-19. The aim is to break the human-to-human transmission chain and allow for targeted public health measures considering pre-symptomatic and asymptomatic transmission possibilities (World Health Organization, 2020a).

Digital tools developed to assist contact tracing vary widely. They include proximity tracing tools, CCTV with facial recognition (FR) and geolocation-quick response code (GEO-QR) tagging systems. Proximity tracing is based on the use of GPS (Silveira, 2021), Bluetooth (Hatke et al., 2020) or ultrasound (Cranor, 2020) technologies that can record movements of individuals and who they have come in contact with. This means that when a person tests positive, people who may have been exposed may be traced, found and notified. The underlying logic is that the risk of exposure depends on the probability of coming into close or frequent contact with the infected person (World Health Organization, 2020b). Several countries (such as China, Russia and South Korea) have utilised facial recognition technology for COVID-19 contact tracing (Ramos, 2020). This level of surveillance requires that the identity of a positive patient is embedded into a biometric database and FR software run over live camera feeds or still images (Berman et al., 2020). This can be used to actively monitor confirmed cases or exposed persons who are self-isolating. QR code scanning technology underpins contact tracing efforts in countries such as Malaysia, Australia and New Zealand (Jahmunah et al., 2021). This requires placing a QR code at a venue and asking people to scan the code with a mobile phone to tag their visit (Nakamoto et al., 2020). Either centralized or decentralized communication protocols shape these digital approaches.

During the early development of CTA's one of the protocols that became popular for systems using centralised servers is the Pan-European Privacy-Preserving Proximity Tracing (PEPP-PT). Although few countries have successfully developed CTAs based on this protocol, in April of 2020 at least eight countries including France, Spain and Germany backed the project developing this protocol. Whilst Germany and Spain pulled out of the PEPP-PT project, France went ahead in developing a CTA called StopCOVID in June 2020 using a variant of PEPP-PT referred to as ROBERT (Robust and Privacy Preserving Proximity Tracing) protocol (O'Brien, 2020). However, France discontinued it a few months later due to a host of problems, including poor download numbers and inefficiency of the app (Schechner, 2020). A revamped version of the app called TousAntiCOVID was launched in October 2020, and by June 2021, it had been downloaded by about 26% of France's population of 67.39 million (World Bank, 2021).

Around the same period (April 2020), the popular protocols for CTAs using decentralised servers were the Decentralised Privacy-Preserving Proximity Tracing (DP-3T) protocol and the Google Apple Exposure Notification System (GAEN). While DP-T3 was developed by an independent group of tech experts based mainly in Europe, GAEN was developed through a collaborative effort of the tech giants Apple and Google, yet it is widely considered a variant of DP-T3.

After Germany and Spain pulled out of the PEPP-PT project, they later opted to use the GAEN API (Application Programming Interface) for their apps. Germany launched its corona-warn app in June 2020 and a year later, it had been downloaded by about 35% of the German population. Conversley, Spain's CTA RadarCOVID was released in August 2020. By June 2021, it had been downloaded by about 15% of the population (RadarCOVID, 2021). In the UK, attempts were made to create a CTA based on proprietary centralised protocols that were developed in-house. This NHSx app was discontinued and never launched for public use after trials (including one on the Isle of Wight in March and April 2020) showed that the app was highly inefficient and unpopular due to several issues, including privacy concerns (White, 2021). A separate version called NHS COVID-19 app was developed with the GAEN system and launched in England and Wales in September 2020 has been downloaded by 43.37% of the population (NHS, 2021). Table 4.1 below provides an overview of some of these developments.

Table 4.1 Comparison of population and download statistics

Country	Population (Mio)	CTA	Launch date	Protocol	Downloads (Mio) (June 2020)	Downloads (Mio) (June 2021)	% of Population
France	67.39	TousAntiCovid	22-Oct-20	ROBERT	x	16.5	26.02
		StopCovid	02-Jun-20	ROBERT	1	x	1.48
Germany	83.24	Corona-warn app	16-Jun-20	GAEN	10	29.2	35.07
Spain	47.35	Radar Covid	10-Aug-20	GAEN	x	7.2	15.20
UK[a]	59.72	NHS COVID-19	24-Sep-20	GAEN	x	26.1	43.70

[a]Note that the UK population referred to here is for England and Wales

4.4   Methodology

To enable a detailed understanding of the issues surrounding the development and use of CTAs, a Multivocal Literature Review (MLR) approach was applied. MLR is a systematic literature review that includes grey literature (GL) alongside peer-reviewed articles (Garousi et al., 2019). Ogawa and Malen (1991), who developed this methodology, describe multivocal literature as "all accessible writings on a common, often contemporary topic" which embody the voices or views of a diverse set of authors, including academics, practitioners, journalists, policy centres, independent research and development firms, state offices etc.

MLR has been utilised in a variety of fields, including software engineering (Garousi et al., 2019), education (Ogawa & Malen, 1991), management (Adams et al., 2017), finance, and health science (Saleh et al., 2014; Tarhan et al., 2020). Yet its application in Information Systems (IS) research is relatively new. The contemporary nature of many IS studies and the growing use of grey literature as a means for communication and dissemination means that other forms of systematic literature review underutilise this valuable source of information. By applying the MLR approach, this chapter also seeks to take advantage of the diversity of material produced outside the academic peer-reviewed process. Furthermore, the emerging nature of the COVID-19 health crisis means that adequate, relevant data may be unavailable for this study if traditional data sources are relied upon.

However, there are challenges in dealing with grey literature that must be acknowledged, including lacking an extensive peer-review process like scientific publications, limitations in scientific rigour, and limited methodological descriptions in grey literature that enable an evaluation of the quality of the research process (Adams et al., 2017). Given these challenges, Garousi et al. (2019) developed a Taxonomy for multivocal literature designed to minimise these issues by recognising four categories of literature based on the expertise involved, credibility, and publisher control (see Fig. 4.1).

Figure 4.1 shows a spectrum of four colours of increasing darkness plotted on two axes representing outlet control and source expertise. Outlet control is described as the extent of moderation or conformance with explicit and transparent knowledge creation criteria. Source expertise is the extent to which the authority of the content producer can be determined and is a measure of the author's credibility (Adams et al., 2017). Based on these dimensions, peer-reviewed journals are represented in white, with increasing tiers representing ever lower outlet control and credibility.

These findings are primarily based on 'white literature' and include tier 1 and tier 2 grey literature. The procedure followed is primarily based on the guidelines for MLR developed by Garousi et al. (2019), itself adapted from the guidelines for systematic literature reviews provided by Kitchenham and Charters (2007). It includes specifying the research question(s), developing and evaluating the review protocol, search process and source selection, study quality assessment, data extraction, and data synthesis. One of the ways that MLR was utilised for this chapter was

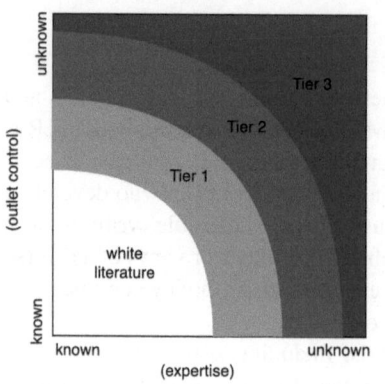

Tier 3 Grey literature:
low outlet control & credibility
Includes tweets, non-institutional blogs, Wikipedia articles

Tier 2 Grey literature:
moderate outlet control & credibility
Includes institutional reports, news articles, institutional blogs

Tier 1 Grey literature:
significant outlet control & credibility
Includes books, book chapters, government report, white papers

Fig. 4.1 Classes of multivocal literature (adapted from Garousi et al., 2019)

to understand the gaps in the literature on doing RRI in a health crisis. The research question identified in this case was 'what is known from the existing literature about doing RRI in a health crisis?'

This was an important question because the nature of health crises means that rapid solutions are sought and often involve the development and deployment of new digital technologies. Stahl (2020) argues that although these technologies are usually well-intentioned, they are generally potentially problematic, and RRI may provide a valuable approach for addressing such problems. Thus, a literature search was carried out on the Scopus abstract and citation database and Google. These databases were selected because of their size, scope, user-friendliness, search simplicity, and institutional support for Scopus. The search strategy used to determine what the existing literature says about RRI in a health crisis used the keywords "Responsible Research and Innovation OR RRI and crisis" covering the period 2011 to 2021. The literature search goes back to 2011 (before COVID-19 developed in 2019) to consider relevant RRI years and other crises during this period, e.g. ZIKA Virus (Díaz-Menéndez & Crespillo-Andújar, 2017), MERS (WHO, 2015) and EBOLA (Quaglio et al., 2016). Articles without reference to the relevant themes in their titles or abstracts were eliminated, and the remaining were read in full to capture the full scope.

Of the 50 articles that resulted from this strategy, 6 articles in Scopus were found to have some relationship to the question of existing literature on RRI in crisis. After eliminating duplicates, 1 additional article was found via Google. However, three of these articles (Carrier & Irzik, 2019; Stilgoe et al., 2013; Buscher et al., 2018) focused on subjects unrelated to health crises. As outlined earlier, the strategy for this chapter also includes determining how four European countries have responded to the development of CTAs to capture the interaction of factors and events. The four countries are Germany, Spain, the UK, and France.

The countries were selected because they are the biggest funders of the Horizon 2020 Framework Work Programme, which is the EU primary mechanism for funding RRI related activities in Europe. RRI is a cross-cutting issue in Horizon 2020.

Therefore, the most prominent funders (who are also amongst the wealthiest countries in Europe) are likely to have the highest capacity for RRI related activities. According to the European Commission (2020), these countries rank the top four based on contribution rates and participation in Horizon 2020. A similar search strategy was used in both Google and Scopus. The terms used were 'country name OR adjective' AND COVID-19 AND "Contact Tracing app". For example, for Spain, the search term was Spain OR Spanish AND COVID-19 AND "Contact Tracing app". A total of 80 relevant articles were found; grey literature constituted 26, the others were journal articles. The findings provide an understanding of the activities of four countries during the development and deployment of CTAs that can be classified using the dimensions of RRI.

4.5   Elements of RRI in the Development of Contact Tracing Apps

To understand the activities that can be mapped to RRI during the COVID-19 pandemic, three dimensions of RRI (anticipation, reflexivity, and inclusive public participation) were considered. Anticipation is a key dimension of RRI that Stilgoe et al. (2013) argue involves systematically thinking of opportunities to develop socially robust research. This requires researchers and organisations to consider what is known, likely, and plausible. Furthermore, they maintain that 'reflexivity at the level of institutional practice means holding a mirror up to one's activities, commitments and assumptions, being aware of limits of knowledge and being mindful that a particular framing of an issue may not be universally held'. Finally, they argue that researchers and organisations must move beyond engagement with stakeholders to include the broader public for inclusive participation, which they also link to the provision of clear communication of the nature and purpose of the project and mechanisms for understanding public and stakeholder views.

Although there was no explicit evidence found that shows the RRI was addressed during the development of the CTAs in France, the UK, Spain, and Germany, the literature review has identified some interesting linkages to this framework. The following highlights the findings of the literature review for the four countries:

France
In France there appears to have been some reflexivity at the governmental level during the development of the StopCOVID app. An instance of this is seen in the Government's request for a debate on the development and deployment of the app in parliament and to seek the legal advice of the National Commission on Informatics and Liberty - CNIL (CNIL, 2020; Rowe et al., 2020). CNIL is an independent data protection body set up to ensure that data privacy laws are maintained in collecting and using personal data.

The French government also asked the National Pilot Committee for Digital Ethics to reflect on the CTA and issues around digital ethics during the health crisis

(National Pilot Committee for Digital Ethics, 2020; Institut Français des Droits et Libertés, 2020). While the government was responsive to most of the advice from these bodies, CNIL (2020) highlights that one issue that the government didn't anticipate was the generation of false positives due to failure to take the context of contacts being made into account, e.g. the type of protective equipment an individual might have. Rowe et al. (2020) have pointed out other government failings in this regard, maintaining that interventions like the CTA were developed based on incomplete knowledge, there was a lack of readiness for the crisis, and they defaulted to a strategy that relied on executive summaries from the UK's NHS rather than anticipatory research.

At the development stage of the app, some effort was put towards being inclusive and participatory. The StopCOVID development team comprised several private companies and public institutions (Institut Français des Droits et Libertés, 2020). Nevertheless, Information Systems research findings were not considered in the app's planning, design, and deployment, and crucial information regarding data collection, privacy, security, and data processing, storage, and reuse was not explained clearly to the public (Rowe et al., 2020). It appears that as many citizens had to rely on other sources of information, their trust in the system waned and acceptance levels for the app dropped from 80% to 44% between March and April 2020 (Guillon & Kergall, 2020). Also, Montagni et al. (2020) who explored reasons for the low uptake of contact tracing apps among university students in the health disciplines, found a limited awareness and a considerable amount of misinformation about the app among this group. These issues raise questions about inclusivity and the participation of the public during the development and deployment of the CTA.

The United Kingdom

In the UK, a similar situation can be seen for the RRI dimensions of anticipation, inclusive participation, and reflexivity. In terms of anticipation, the government anticipated some of the privacy issues that could result from the use of location data from mobile network operators to reveal trends in social mobility, even if such data were aggregated and anonymised. This is because the UK is among the few countries in Europe that decided against collecting and submitting such data to the EC's Joint Research Centre (European Commission, 2011). Like Denmark, concerns focused on the reversibility of 'anonymised' data and potential third-party access. Nevertheless, whilst some level of anticipation of the effect of CTAs on different demographics was considered, Guinchard (2021) argues that the timing of development in the middle of the pandemic raises important questions as to why no consideration was made to develop such apps years earlier.

Ryder et al. (2020) maintain that the UK government engaged with several groups and organisations during the development of CTAs in terms of inclusive participation. These organisations include (i) the National Health Service (NHS); (ii) the Information Commissioner's Office (ICO); (iii) the National Data Guardian's Panel; and (iv) patient advocacy groups like 'Understanding Patient Data'. However, it appears that little dialogue happened early enough, and insufficient effort was put into public communication as the information provided was 'scattered and vague' and did not help allay concerns of the problematic impact of the app (Guinchard, 2021; Williams et al., 2021).

Arguing for clear public communication, point out the need for any messaging around the app to be done in such a way as to alleviate fears about surveillance, hacking and to reduce anxiety around the epidemic. Also, McGregor et al. (2020) point out that insufficient information was provided about the operation of the app and its data flows, the legal basis, oversight, accountability, possible future uses of data and impact on human rights, as well as remedies. These issues indicate a disconnect between the development of the app and public engagement, and consequences included lack of information, increasing mistrust of the app, and growing unwillingness to download and use it (Ada Lovelace Institute, 2020b).

After much public backlash, the government showed some reflexivity by agreeing to open up the code used to develop its CTA to the public and published a Data Protection Impact Assessment (DPIA)(Ryder et al., 2020; Guinchard, 2021). However, further issues, including poor data security and privacy controls (Culnane & Teague, 2020), meant that the UK government discontinued the development of a centralised CTA (the NHSx app) and opted instead for the GAEN system (Wise, 2020; Ada Lovelace Institute, 2020a; French et al., 2020). Once again, reflexivity in these cases appears to have been an afterthought predicated on extensive criticism, as many remedial actions could have been taken sooner.

Spain

The publications reviewed convey an impression that Spain has a more limited experience of the activities that can be mapped to RRI than the other countries examined. Although some interesting points are highlighted here, it was challenging to find relevant publications based on the search criteria used and map them to the government's actions on the RRI dimensions. For anticipation, like the UK, the timing of the app's development in Spain has been criticised by Zeng et al. (2020), who compare it to countries like Singapore which quickly saw the potential of CTAs and developed them early in the pandemic. One of the press conferences given by Spain's Interior Minister Grande-Marlaska, who argued for the use of geolocation of citizens' mobile phones not just for contact tracing but also for policing (Binnie, 2020), shows another poor example of anticipation. This would have been a problematic departure point from previous use of geolocation data in Spain when anonymised, and aggregated mobile phone location data was used to track people's movement to determine compliance to lockdown rules (Rodriguez-Ferrand, 2020). The government likely wanted to further capitalise on an earlier poll that showed 47% of citizens were willing to share personal information to contain the pandemic (Miláns del Bosch, 2020), but this plan quickly changed public perception.

That the Spanish government decided against using geolocation for policing is an example of reflexivity as they likely realised that their framing of the issues was not universally accepted. Nevertheless, Hernández-Quevedo et al. (2020) point out that as countries like Spain continue to struggle with the difficult balance between effective contact tracing and privacy preservation, there is a need for greater transparency in the collection and use of data to ensure that privacy is prioritised. Furthermore, transparency is closely linked to openness, and this, in turn, may be linked to inclusive public participation. In this regard, Weiß et al. (2021) point out that the Spanish CTA has an open-source repository for its code that acts as a dedicated information hub. Similarly, Raman et al. (2021) suggest that in terms of factors that determine the effectiveness of CTAs, Spain appears to have done quite well in the areas of accessibility and raising awareness.

Germany

Germany appears to have fared little better than France and the UK regarding the activities identified using the outlined RRI components. One example is the push to utilise the PEPP-PT protocol to facilitate digital contact tracing (Walther et al., 2020; Moreno, 2020). This led to intense public backlash (Leith & Farrell, 2020), which may be interpreted as poor anticipation by the government as it failed to consider the societal desirability of such a centralised system especially considering historical issues around surveillance in Germany (Eley, 2016; Schaer, 2019). The government also appears not to have anticipated issues with the use of the app in public transport systems; Leith and Farrell (2020) demonstrate that the apps are ineffective in trams, likely due to the reflection of radio signals from the metal structure. Grill et al. (2021) assessed sociodemographic characteristics of users of CTAs in Germany and found that on the one hand, users of the app were less likely to be female, younger, and to have a lower family income, but on the other, they were more likely to live in one of the Western federal states. This suggests that the government has inadequately anticipated how factors like education, income and region affect usage despite previous studies identifying such problems (McAuley, 2014; Carroll et al., 2017; Latulippe et al., 2017).

This is also significant for inclusive public participation as it raises important questions about the German government's prioritisation of public engagement during the development of the app (Zimmermann et al., 2021). It must, however, be noted that although the population download rates of the CTA in Germany remains relatively low (Amann et al., 2021; Zimmermann et al., 2021; Blom et al., 2021), Munzert et al. (2021) suggest that considerable awareness of the app was generated and the provision of monetary incentives for downloading the app might be more effective than further awareness-raising. Also, the government has been hailed for its open-source approach, which enables public scrutiny of the apps source code and increased transparency (Sonnekalb et al., 2020; Amann et al., 2021; Weiß et al., 2021). However, Grill et al. (2021) argue there has been a 'missed communication opportunity' because many non-users are not aware of the usefulness and effectiveness of the app, and, the government has been criticised for the lack of transparency and clear communication about its purpose and function (Amann et al., 2021). Public outreach by political representatives has been particularly problematic and has created some confusion (Ranisch et al., 2020); in March 2020, Health Minister Jens Spahn commented that the government was trying to extend the German Epidemic Law to enable tracking and surveillance, sparking intense criticism.

One of the best indications of reflexivity on the part of the German Government during the development of its CTA can be seen in its decision to adopt the GAEN Framework despite being one of the biggest state supporters of the PEPP-PT protocol (Walther et al., 2020; Moreno, 2020). Interestingly, this change was precipitated by privacy concerns (Reintjes, 2020), massive criticism (Ranisch et al., 2020), considerable indignation (Grill et al., 2021) and enormous outcry from academics and organisations (Bagchi et al., 2020). Despite this being an example of some reflexivity, it is noteworthy that the decision to change course has been widely criticised, with some questioning the trustworthiness of big tech companies, and others perceiving this as another example of big tech dominance (Amann et al., 2021).

All these paint an interesting picture of activities that have some resemblance to RRI during the development of CTAs. Although few and far between, instances of such activities have been described here, along with the highlighting of situations where there appears to have been failings. The issues identified are illustrated in Table 4.2.

4.6 Implications for RRI in Health Crisis Situations

Despite there being no explicit mention of RRI during the development of CTAs in the cases considered here, the analysis has shown that some of the activities during this period can be mapped to RRI. Using RRI as an analytic tool, the chapter has also identified and classified key issues in the process of development of CTAs. For example, none of the governments had anticipated the need for CTAs and was unprepared to rapidly develop and deploy them. In many cases, they did not anticipate societal concerns like those related to privacy and trust. There were also issues with inclusive participation as many felt information about CTAs was poorly communicated and inadequate, and there were problems with public outreach and transparency. Likewise, reflexivity on the governments' part appears to be mainly due to intense public criticism and backlash and little to do with 'holding a mirror up to their own activities, commitments and assumptions (Stilgoe et al., 2013)'.

Thus, this chapter has identified issues that could be addressed with the application of RRI during the COVID-19 health crisis. This chapter has also indicated that although little appears to have been said about the opportunities for RRI, an important theme in the discourse on RRI for crisis management is inclusive engagement. This chapter has demonstrated how all previous literature on RRI in crises either called for greater participation or started initiatives to advance participation. Issues around inclusive participation have also featured in the development of CTAs and, along with the other issues identified, may have resulted in low public acceptance of COVID-19 CTAs.

Considering that crises management requires speed (Nickson et al., 2020; Am et al., 2020), the possibility for effectively engaging in inclusive public participation must be questioned, as indeed for other RRI dimensions like anticipation and reflexivity. Anticipatory techniques like foresight, horizon scanning and technology

Table 4.2 Summary of identified issues classified using RRI dimensions

	France	UK	Spain	Germany
Anticipation	The possibility of generating numerous false positives was not anticipated by the government	The timing of development in the middle of the pandemic raises important questions as to why no consideration was made to develop such apps years earlier	The timing of the development of the app in Spain has been criticised	Poor anticipation of the societal desirability of a centralised system especially considering issues around surveillance in the country's history
	The contact tracing app was developed based on incomplete knowledge and there was a lack of readiness for the crisis		Poor anticipation of societal desirability for the use of geolocation of citizens' mobile phones not just for contact tracing, but also for policing	Poor anticipation of the level of inefficiency of the app in public transport systems
				The government also appears to have inadequately anticipated how factors like education, income and region affect usage
Inclusive participation	Information Systems research findings were not considered in the planning, design and deployment of the app	The dialogue did not happen early enough and insufficient effort was put into public communication as the information provided was 'scattered and vague' and did not help allay concerns of the problematic impact of the app	The need for greater transparency in the collection and use of data to ensure that privacy is prioritised	Important questions about how the German governments prioritised public engagement during the development of the app

(continued)

Table 4.2 (continued)

	France	UK	Spain	Germany
	Crucial information regarding data collection, privacy, security, as well as data processing, storage, and reuse were not explained in plain language to the public	Messaging around the app must be done in such a way as to alleviate fears about surveillance, hacking, and to reduce anxiety around the epidemic		Missed communication opportunities as considerable part of non-users remain unaware of the usefulness and effectiveness of the app
		Insufficient information was provided about the operation of the app and its data flows, the legal basis, oversight, accountability, possible future uses of data and impact on human rights, as well as remedies		Lack of transparency and clear communication about the purpose and function of the app
				Problems with public outreach by political representatives which sometimes create confusion
Reflexivity		Evidence of reflexivity only as a result of intense public backlash		Reflexivity only as result of massive criticism considerable indignation and huge outcry from academics and organisations

assessment used for looking ahead at the societal impact of technology often involve prolonged periods of assessments and deliberations that are unsuitable in a crisis. This is equally true for institutional reflexivity mechanisms such as codes of conduct, moratoriums, and standards adoption. Despite these challenges, RRI has its uses in a crisis. It has been shown here how it can be used effectively as an analytical tool to identify opportunities for improving techno-social responses to crises and for reflection on the development of emerging technologies in a situation like those created by the COVID-19 pandemic.

Acknowledgements This research has received funding from the European Union's Horizon 2020 Framework Programme for Research and Innovation under the Grant Agreements No. 720270 (HBP SGA1), 785907 (HBP SGA2), 945539 (HBP SGA3) and the Framework Partnership Agreement No. 650003.

References

Ada Lovelace Institute. (2020a) *Confidence in a crisis? Building public trust in a contact tracing app.* Available from: https://www.adalovelaceinstitute.org/report/confidence-in-crisis-building-public-trust-contact-tracing-app/. Accessed 19 May 2021.

Ada Lovelace Institute. (2020b) *No green lights, no red lines.* Available from: https://www.adalovelaceinstitute.org/report/covid-19-no-green-lights-no-red-lines/. Accessed 19 May 2021.

Adams, R. J., Smart, P., & Huff, A. S. (2017). Shades of grey: Guidelines for working with the Grey literature in systematic reviews for management and organizational studies: Shades of grey. *International Journal of Management Reviews, 19*(4), 432–454. https://doi.org/10.1111/ijmr.12102

Am, J. B., Furstenthal, L., Jorge, F., & Roth, E. (2020). *Innovation in a crisis: Why it is more critical than ever | McKinsey.* McKinsey and Company.

Amann, J., Sleigh, J., & Vayena, E. (2021). Digital contact-tracing during the Covid-19 pandemic: An analysis of newspaper coverage in Germany, Austria, and Switzerland. *PLoS One, 16*(2), e0246524. https://doi.org/10.1371/journal.pone.0246524

Bagchi, K. K. J., et al. (2020). *Digital tools for COVID-19 contact tracing: Identifying and mitigating the equity, privacy, and civil liberties concerns.* Edmond J. Safra Center for Ethics.

Berman, G., et al. (2020). *Digital contact tracing and surveillance during COVID-19. General and child-specific ethical issues.* UNICEF Office of Research.

Binnie, I. (2020). Spain's COVID tracing app tries to balance public health with privacy. *Reuters.*

Blom, A. G., et al. (2021). Barriers to the large-scale adoption of a COVID-19 contact tracing app in Germany: Survey study. *Journal of Medical Internet Research, 23*(3), e23362. https://doi.org/10.2196/23362

Braun, R., et al. (2020). COVID-19 and the onlineification of research: Kick-starting a dialogue on Responsible online Research and Innovation (RoRI). *Journal of Responsible Innovation, 7*(3), 680–688. https://doi.org/10.1080/23299460.2020.1789387

Buscher, M., et al. (2018). The IsITethical? Exchange: Responsible research and innovation for disaster risk management. In K. Boersma & B. Tomaszewski (Eds.), *Proceedings of the 15th ISCRAM Conference* (pp. 254–267). 15th International Conference on Information Systems for Crisis Response and Management.

Carrier, M., & Irzik, G. (2019). Responsible research and innovation: Coming to grips with an ambitious concept. *Synthese.* https://doi.org/10.1007/s11229-019-02319-1. Accessed 02 July 2020.

Carroll, J. K., et al. (2017). Who uses mobile phone health apps and does use matter? A secondary data analytics approach. *Journal of Medical Internet Research, 19*(4), e125. https://doi.org/10.2196/jmir.5604

Cave, S., et al. (2021). Using AI ethically to tackle covid-19. *BMJ, 372*, n364.

CNIL. (2020). *Deliberation N° 2020-056 from 25 May 2020 delivering an opinion on a draft decree relating to the mobile application known as 'StopCovid'*. Commission nationale de l'informatique et des libertés.

Cranor, L. F. (2020). Digital contact tracing may protect privacy, but it is unlikely to stop the pandemic. *Communications of the ACM, 63*(11), 22–24. https://doi.org/10.1145/3423993

Crook, P., et al. (2017). Lack of secondary transmission of Ebola virus from healthcare worker to 238 contacts, United Kingdom, December 2014. *Emerging Infectious Diseases, 23*(12), 2081–2084. https://doi.org/10.3201/eid2312.171100

Culnane, C., & Teague, V. (2020). Security analysis of the NHS COVID-19 App. [Online] *State of IT*. Available from: https://stateofit.com/UKContactTracing/. Accessed 22 May 2021.

Díaz-Menéndez, M., & Crespillo-Andújar, C. (2017). *Zika virus infection: Risk of spreading in Europe*. Springer International Publishing.

Eley, G. (2016). Surveillance in German History. *German History, 34*(2), 293–314.

European Commission. (2011). *Proposal for a regulation of the European parliament and of the council establishing horizon 2020 – The framework programme for research and innovation (2014–2020)*. European Commission.

European Commission. (2015). *Indicators for promoting and monitoring responsible research and innovation: Report from the Expert Group on policy indicators for responsible research and innovation*. LU. Publications Office.

European Commission. (2020). *Horizon 2020 country profiles: Key, research-related data and recent project success stories*. Available from: https://ec.europa.eu/research/horizon2020/index.cfm?pg=country-profiles. Accessed 27 November 2020.

French, M., et al. (2020). Corporate contact tracing as a pandemic response. *Critical Public Health, 32*(1), 48–55. https://doi.org/10.1080/09581596.2020.1829549

Garousi, V., Felderer, M., & Mäntylä, M. V. (2019). Guidelines for including grey literature and conducting multivocal literature reviews in software engineering. *Information and Software Technology, 106*, 101–121.

Grasenick, K., & Guerrero, M. (2020). Responsible research and innovation & digital inclusiveness during Covid-19 crisis in the human brain project (HBP). *Journal of Responsible Technology, 1*, 100001. https://doi.org/10.1016/j.jrt.2020.06.001

Grill, E., et al. (2021). Sociodemographic characteristics determine download and use of a Corona contact tracing app in Germany – Results of the COSMO surveys. *PsychArchives, 16*(9), e0256660. https://doi.org/10.23668/psycharchives.4713

Guillon, M., & Kergall, P. (2020). Attitudes and opinions on quarantine and support for a contact-tracing application in France during the COVID-19 outbreak. *Public Health, 188*, 21–31. https://doi.org/10.1016/j.puhe.2020.08.026

Guinchard, A. (2021). Our digital footprint under Covid-19: Should we fear the UK digital contact tracing app? *International Review of Law, Computers & Technology, 35*(1), 84–97. https://doi.org/10.1080/13600869.2020.1794569

Hajer, M. (2003). Policy without polity? Policy analysis and the institutional void. *Policy Sciences, 36*(2), 175–195. https://doi.org/10.1023/A:1024834510939

Hatke, G. F., et al. (2020). *Using bluetooth low energy (BLE) signal strength estimation to facilitate contact tracing for COVID-19*. Masschusetts Institute of Technology.

Hernández-Quevedo, C., et al. (2020). Effective contact tracing and the role of apps: Lessons from Europe. *Eurohealth, 26*(2), 5.

Miláns del Bosch, L. (2020). Spain suggests COVID-19 is changing views on data-sharing. [Online] *Oliver Wyman Forum*. Available from: https://www.oliverwymanforum.com/future-of-data/2020/apr/spain-suggests-covid-19-is-changing-views-on-data-sharing.html. Accessed 17 June 2021.

Hogben, M., et al. (2016). Partner services in sexually transmitted disease prevention programs: A review. *Sexually Transmitted Diseases, 43*(2S), S53–S62. https://doi.org/10.1097/OLQ.0000000000000328

Institut Français des Droits et Libertés. (2020). *Coronavirus pandemic in the EU – Fundamental Rights Implications.* European Union Agency for Fundamental Rights (FRA).

Jahmunah, V., et al. (2021). COVID-19 contact tracing and prediction: A review of the state-of-the-science. *International Journal of Imaging Systems and Technology, 31*(2), 455–471. https://doi.org/10.1002/ima.22552

Khallouf, A. B. (2018). Humanitarian Medical Cloud Computing Systems (HMCCS): Applying RRI Framework in the Emergency Preparedness and Response to the Public Health Catastrophe Triggered by the Syrian Crisis. In *Responsible research and innovation actions in science education, gender and ethics: Cases and experiences* (pp. 39–46). Springer.

Kitchenham, B. A., & Charters, S. M. (2007). *Guidelines for performing literature review in software engineering.* EBSE Technical Report.

Kwok, K. O., et al. (2019). Epidemic models of contact tracing: Systematic review of transmission studies of severe acute respiratory syndrome and middle east respiratory syndrome. *Computational and Structural Biotechnology Journal, 17*, 186–194. https://doi.org/10.1016/j.csbj.2019.01.003

Latulippe, K., Hamel, C., & Giroux, D. (2017). Social health inequalities and eHealth: A literature review with qualitative synthesis of theoretical and empirical studies. *Journal of Medical Internet Research, 19*(4), e136. https://doi.org/10.2196/jmir.6731

Leith, D. J., & Farrell, S. (2020). Measurement-based evaluation of Google/Apple Exposure Notification API for proximity detection in a light-rail tram Soldani, J. (ed.). *PLoS One, 15*(9), e0239943. https://doi.org/10.1371/journal.pone.0239943

McAuley, A. (2014). Digital health interventions: Widening access or widening inequalities? *Public Health, 128*(12), 1118–1120. https://doi.org/10.1016/j.puhe.2014.10.008

McGregor, L. et al. (2020). *Written evidence from professor Lorna McGregor et al (COV0090).* Available from: https://committees.parliament.uk/writtenevidence/3973/html. Accessed 19 May 2021.

Montagni, I., et al. (2020). The French Covid-19 contact tracing app: Knowledge, attitudes, beliefs and practices of students in the health domain. *Medical Education*, online first.

Monteiro, M., Shelley-Egan, C., & Dratwa, J. (2017). On irresponsibility in times of crisis: Learning from the response to the Zika virus outbreak. *Journal of Responsible Innovation, 4*(1), 71–77. https://doi.org/10.1080/23299460.2017.1312959

Moor, J. H. (1985). What is computer ethics?*. *Metaphilosophy, 16*(4), 266–275.

Moor, J. H. (2005). Why we need better ethics for emerging technologies. *Ethics and Information Technology, 7*(3), 111–119. https://doi.org/10.1007/s10676-006-0008-0

Moreno, N. (2020). *Coronavirus and contact tracing apps: Global privacy perspective.* Addleshaw Goddard LLP.

Morley, J., et al. (2020). Ethical guidelines for COVID-19 tracing apps. *Nature, 582*(7810), 29–31. https://doi.org/10.1038/d41586-020-01578-0

Munzert, S., et al. (2021). Tracking and promoting the usage of a COVID-19 contact tracing app. *Nature Human Behaviour, 5*(2), 247–255. https://doi.org/10.1038/s41562-020-01044-x

Nakamoto, I., et al. (2020). A QR code–based contact tracing framework for sustainable containment of COVID-19: Evaluation of an approach to assist the return to normal activity. *JMIR mHealth and uHealth, 8*(9), e22321. https://doi.org/10.2196/22321

National Pilot Committee for Digital Ethics. (2020). *Reflections and warning points on digital ethics issues in situations of acute health crisis.* Comité Consultatif National d'Éthique CCNE.

NHS. (2021). *NHS COVID-19 app statistics.* Available from: https://stats.app.covid19.nhs.uk/. Accessed 16 July 2021.

Nickson, S., Thomas, A., & Mullens-Burgess, E. (2020). *Decision making in a crisis: First responses to the coronavirus pandemic.* Institute for Government.

Nuffield Council on Bioethics. (2020). Twenty questions about the NHSX contact tracing app. [Online] *The Nuffield Council on Bioethics*. Available from: https://www.nuffieldbioethics.org/blog/twenty-questions-about-the-nhsx-contact-tracing-app. Accessed 13 May 2020.

O'Brien, C. (2020). France offers a case study in the battle between privacy and coronavirus tracing apps. [Online] *VentureBeat*. Available from: https://venturebeat.com/2020/05/18/france-offers-a-case-study-in-the-battle-between-privacy-and-coronavirus-tracking-apps/. Accessed 16 July 2021.

Ogawa, R. T., & Malen, B. (1991). Towards rigor in reviews of multivocal literature: Applying the exploratory case study method. *Review of Educational Research, 61*(3), 265–286. https://doi.org/10.3102/00346543061003265

Pitlik, S. D. (2020). COVID-19 compared to other pandemic diseases. *Rambam Maimonides Medical Journal, 11*(3), e0027. https://doi.org/10.5041/RMMJ.10418

Quaglio, G., et al. (2016). Ebola: Lessons learned and future challenges for Europe. *The Lancet Infectious Diseases, 16*(2), 259–263. https://doi.org/10.1016/S1473-3099(15)00361-8

RadarCOVID. (2021). *App RadarCOVID statistics*. Available from: https://radarcovid.gob.es/estadisticas/codigos-solicitados-a-casos-confirmados. Accessed 16 July 2021.

Raman, R., et al. (2021). COVIDTAS COVID-19 tracing app scale—An evaluation framework. *Sustainability, 13*(5), 2912. https://doi.org/10.3390/su13052912

Ramos, L. F. M. (2020). Evaluating privacy during the COVID-19 public health emergency: The case of facial recognition technologies. In *Proceedings of the 13th International Conference on Theory and Practice of Electronic Governance. ICEGOV 2020: 13th International Conference on Theory and Practice of Electronic Governance* (pp. 176–179). ACM.

Ranisch, R., et al. (2020). Ethics of digital contact tracing apps for the Covid-19 pandemic response. *Kompetenznetz Public Health COVID, 19*(10), 13140. https://doi.org/10.13140/RG.2.2.23149.00485

Reintjes, R. (2020). Lessons in contact tracing from Germany. *BMJ, 369*, m2522. https://doi.org/10.1136/bmj.m2522

Riley, S., et al. (2003). Transmission dynamics of the etiological agent of SARS in Hong Kong: Impact of public health interventions. *Science, 300*(5627), 1961–1966. https://doi.org/10.1126/science.1086478

Rodriguez-Ferrand, G. (2020). *Regulating electronic means to fight the spread of COVID-19: Spain*. The Law Library of Congress.

Rowe, F., Ngwenyama, O., & Richet, J.-L. (2020). Contact tracing apps and alienation in the age of COVID-19. *European Journal of Information Systems, 29*(5), 545–562. https://doi.org/10.1080/0960085X.2020.1803155

Ryder, M., et al. (2020). *COVID-19 & Tech responses: Legal opinion*. Matrix Law.

Saleh, A. A., Ratajeski, M. A., & Bertolet, M. (2014). Grey literature searching for health sciences systematic reviews: A prospective study of time spent and resources utilized. *Evidence Based Library and Information Practice, 9*(3), 28–50.

Schaer, C. (2019). Germany's surveillance fears: Thirty years on from the fall of the Berlin wall and the disbanding of the Stasi, Germans worry about who is watching them. *Index on Censorship, 48*(3), 52–53. https://doi.org/10.1177/0306422019875092

Schechner, S. (2020). French contact-tracing app struggles with slow adoption. It isn't alone. *Wall Street Journal*, online first.

Silveira, A. B. (2021). *Digital tracing and COVID-19 – the israeli case*. Institute for Internet and the Just Society.

Sonnekalb, T., et al. (2020). Towards automated, provenance-driven security audit for git-based repositories: Applied to germany's corona-warn-app: Vision paper. In *Proceedings of the 3rd ACM SIGSOFT International Workshop on Software Security from Design to Deployment. SEAD 2020* (pp. 15–18). Association for Computing Machinery. https://doi.org/10.1145/3416507.3423190

Stahl, B. C. (2020). Emerging technologies as the next pandemic? *Ethics and Information Technology, 23*(1), 135–137. https://doi.org/10.1007/s10676-020-09551-1

Sterman, J., & Brauer, A. (2020). Tech expert warns government shouldn't rely on untested apps for contact tracing. [Online] *WSET*. Available from: https://wset.com/news/spotlight-on-america/tech-expert-warns-government-shouldnt-rely-on-untested-apps-for-contact-tracing. Accessed 14 May 2020.

Stilgoe, J., Owen, R., & Macnaghten, P. (2013). Developing a framework for responsible innovation. *Research Policy, 42*(9), 1568–1580. https://doi.org/10.1016/j.respol.2013.05.008

Tarhan, A. K., et al. (2020). Maturity assessment and maturity models in health care: A multivocal literature review. *Digital Health, 6*, 914772. https://doi.org/10.1177/2055207620914772

van der Schaar, M., et al. (2021). How artificial intelligence and machine learning can help healthcare systems respond to COVID-19. *Machine Learning, 110*(1), 1–14. https://doi.org/10.1007/s10994-020-05928-x

Von Schomberg, R. (Ed.). (2011). *Towards responsible research and innovation in the information and communication technologies and security technologies fields*. Publication Office of the European Union.

Walther, M., et al. (2020). *European perspective on tracing tools in the context of Covid-19*. Gibson, Dunn & Crutcher LLP.

Weiß, J.-P., Esdar, M., & Hübner, U. (2021). Analyzing the essential attributes of nationally issued COVID-19 contact tracing apps: Open-source intelligence approach and content analysis. *JMIR mHealth and uHealth, 9*(3), e27232. https://doi.org/10.2196/27232

White, L. (2021). The NHS contact tracing app fell foul of privacy concerns. But did they have the right idea? [Online] *LSE COVID-19*. Available from: https://blogs.lse.ac.uk/covid19/2021/04/21/the-nhs-contact-tracing-app-fell-foul-of-privacy-concerns-but-did-they-have-the-right-idea/. Accessed 16 July 2021.

WHO. (2015). *Middle East respiratory syndrome coronavirus (MERS-CoV) and the risk to Europe*. Available from: https://www.euro.who.int/en/health-topics/health-emergencies/middle-east-respiratory-syndrome-coronavirus-mers-cov-and-the-risk-to-europe. Accessed 19 July 2021.

Williams, S. N., et al. (2021). Public attitudes towards COVID-19 contact tracing apps: A UK-based focus group study. *Health Expectations, 24*(2), 377–385. https://doi.org/10.1111/hex.13179

Wise, J. (2020). Covid-19: UK drops its own contact tracing app to switch to apple and Google model. *BMJ, 369*, m2472. https://doi.org/10.1136/bmj.m2472

World Bank. (2021). *Population, total*. Available from: https://data.worldbank.org/indicator/SP.POP.TOTL. Accessed 16 July 2021.

World Health Organisation WHO. (2014). *Contact tracing during an outbreak of ebola virus disease*. World Health Organization Regional Office for Africa.

World Health Organization. (2020a). *Contact tracing in the context of COVID-19: Interim guidance, 10 May 2020*. World Health Organization.

World Health Organization. (2020b). *Digital tools for COVID-19 contact tracing*. World Health Organization.

Zeng, K., Bernardo, S. N., & Havins, W. E. (2020). The use of digital tools to mitigate the COVID-19 pandemic: Comparative retrospective study of six countries. *JMIR Public Health and Surveillance, 6*(4), e24598. https://doi.org/10.2196/24598

Zimmermann, B. M., et al. (2021). Early perceptions of COVID-19 contact tracing apps in German-speaking countries: Comparative mixed methods study. *Journal of Medical Internet Research, 23*(2), 1–17. https://doi.org/10.2196/25525

Open Access This chapter is licensed under the terms of the Creative Commons Attribution 4.0 International License (http://creativecommons.org/licenses/by/4.0/), which permits use, sharing, adaptation, distribution and reproduction in any medium or format, as long as you give appropriate credit to the original author(s) and the source, provide a link to the Creative Commons license and indicate if changes were made.

The images or other third party material in this chapter are included in the chapter's Creative Commons license, unless indicated otherwise in a credit line to the material. If material is not included in the chapter's Creative Commons license and your intended use is not permitted by statutory regulation or exceeds the permitted use, you will need to obtain permission directly from the copyright holder.

Chapter 5
Uncertainty, Vaccination, and the Duties of Liberal States

Pei-Hua Huang

5.1 Introduction

The highly contagious and fast-evolving COVID-19 virus prompted governments worldwide to take unprecedented emergent measures to contain the pandemic. However, many of these measures give rise to questions regarding the extent to which a liberal state may legitimately intervene in its people's personal decisions in a situation rife with uncertainty. One of the most notable and questionable interventions was the decision to suspend the AstraZeneca vaccine rollout.

The suspension was initially prompted by concerns about exposing people to an undue risk of developing a rare (but severe) cerebral venous thrombosis from the AstraZeneca vaccine. In response to reported cases of this type of thrombosis after receiving the AstraZeneca vaccine, the European Medicines Agencies launched an investigation, with many states suspending their AstraZeneca rollouts. Despite the European Medicines Agency's positive review on the safety of the AstraZeneca vaccine, some states maintained their suspension policy, citing that they had 'better alternatives' for their people (Danish Health Authority, 2021; van Dongen & van Mersbergen, 2021).

Most criticisms of this 'better alternative' account focus primarily on the risks and benefits the prioritisation of other vaccines might bring to society amid a highly time-sensitive battle against COVID-19. These criticisms acknowledge that the countries that suspended the AstraZeneca component of their vaccine rollout had secured more vaccines than they needed, and that these states thus could offer alternative vaccines that were considered safer and more effective. Nevertheless, the suspension may have still caused unnecessary deaths by creating logistical problems and delaying the vaccine rollout.

P.-H. Huang (✉)
Erasmus Medical Centre, Rotterdam, The Netherlands
e-mail: p.huang.1@erasmusmc.nl

© The Author(s) 2022
M. J. Dennis et al. (eds.), *Values for a Post-Pandemic Future*, Philosophy
of Engineering and Technology 40, https://doi.org/10.1007/978-3-031-08424-9_5

Epistemic limitation and uncertainty further complicate the matter of prioritising certain vaccines over others. Due to the urgency of containing the unfolding pandemic, states have had to decide what to do with limited information. While all the vaccines authorised for emergency use have been rigorously tested, given that large-scale vaccination programmes only began in early 2021, it is likely that we will continue to see more rare symptoms identified as the vaccinated population grows (Remmel, 2021). It is also uncertain whether a vaccine that was more effective against the original strain of COVID-19 can continue to outperform other vaccines as new variants continue to emerge. During the composition of this chapter, the newly detected and heavily mutated Omicron variant concerns many medical experts because some of the mutations found in this variant could make the variant more resistant to existing vaccines (Torjesen, 2021).

This uncertainty over emergent effectiveness casts doubt on the legitimacy of the early prioritisation of certain vaccines based on relatively slim margins. In a highly uncertain situation like the COVID-19 pandemic, the epidemiological data changes constantly. An analysis supporting the early prioritisation of a particular vaccine, well supported by the available data at one point in time, may well be undermined as newer data becomes available. Therefore, during periods of uncertainty – periods we may well experience again in our lifetimes – focusing exclusively on risk-benefit analysis provides insufficient normative guidance for public health policymaking.

In this chapter, I use the case of vaccination to develop a duty-based critique. I argue that while a liberal state has a general duty to protect its people's health, the measures this duty can be used to justify are limited. It is especially so when a state tries to use the duty to protect to justify prioritising certain vaccines amidst a highly time-sensitive battle against a pandemic.

Vaccines rely on different technologies, and their mechanisms to trigger an immune response are also different. Because of these differences, each vaccine has different efficacy, side effects, cold-chain requirements, and so forth.[1] It is difficult, if not impossible, to draw a meaningful comparison and conclude which vaccine is ultimately superior. The incommensurability of different kinds of risk also challenges the view that a liberal state may legitimately decide which set of risks one ought to take. The problem of uncertainty also raises questions about whether a state may legitimately appeal to the duty to protect in order to justify vaccine suspension and prioritisation. I argue that when confronted with a highly uncertain situation such as combating a rapidly evolving pandemic, a liberal state must also uphold its duty to properly communicate the known and the unknown to the general public and to assist individuals in determining which risks they are willing to take for their well-being. We can call this duty the duty to facilitate risk-taking.

---

[1] For instance, the Pfizer-BioNTech vaccine must be stored in a specially designed refrigerator at an extremely low temperature (−80 °C to −60 °C) while the AstraZeneca vaccine can be stored in an ordinary refrigerator between 2 °C and 8 °C (National Health Service, 2021). For a quick comparison of the major COVID-19 vaccines, see Ketella (2021).

5.2 A Background: The Better Alternatives

COVID-19 vaccines rely on different technology platforms to trigger an immune response (Katella, 2021). There are at least nine different technology platforms under research and development (Le et al., 2020). Currently, the most widely used vaccines are based on the following technologies: messenger RNA (Pfizer-BioNTech, Moderna), adenovirus vector (AstraZeneca, Sputnik V, Johnson & Johnson), and inactive virus (SinoVac). In addition, several vaccines developed with other technologies like protein subunit, virus-like particles, and DNA have entered Phase II/III clinical trials as of late 2021.[2]

Because of these differences, the mechanism to activate immunity against COVID-19 varies from vaccine to vaccine. For instance, a messenger RNA-based vaccine builds up immunity by producing a coronavirus spike protein and using the protein to teach the body to identify and destroy the virus. Conversely, vaccines based on adenovirus vector technology use modified adenoviruses to trigger a systemic immune response.

The decision to prioritise certain vaccines over others was based mainly on considerations of efficacy against COVID-19. Since COVID-19 vaccines utilise different technologies, it should not be surprising that some vaccines are more effective at protecting people from contracting COVID-19. According to the information provided by the World Health Organisation, the Pfizer-BioNTech vaccines and the Moderna vaccines' efficacy against the original strain of COVID-19 are at the top, at 95% and 94%, respectively (Baden et al., 2021; Polack et al., 2020). Conversely, while still providing sufficient protection (60–70%), the efficacy against symptomatic COVID-19 of the Johnson & Johnson vaccine and the AstraZeneca vaccine is relatively low compared to the two messenger RNA vaccines (Sadoff et al., 2021; Voysey et al., 2021).

It is understandable that certain states decided not to resume the rollout of the AstraZeneca vaccine even after the European Medicines Agency's investigation showed that the benefits of receiving the AstraZeneca vaccine significantly outweighed the risk of developing cerebral venous thrombosis. The rationale behind the decision was that a state has a general duty to promote its people's well-being and protect them from undue health risks and other hazards (Daniels, 2017; United Nations, 1948). Therefore, if a state can afford a more effective vaccine against symptomatic COVID-19, it should provide that more effective vaccine.

This duty provides solid ground for governmental interventions in various affairs, including public health policy. For instance, most liberal states have strict regulations for the conduct of clinical trials. The interventions are morally justifiable because they promote the safety and integrity of the research. Moreover, the restrictions help reduce the epistemic cost a person might otherwise need to pay when

---

[2] For the latest information, see the COVID-19 vaccine tracker maintained by the London School of Hygiene and Tropical Medicine: https://vac-lshtm.shinyapps.io/ncov_vaccine_landscape/

deciding whether a clinical trial is worthy of their participation, or which new treatment they would like to receive.

In the case of COVID-19 vaccine development, while research teams received enormous financial and administrative support from the government sector, all vaccines were still subject to rigorous clinical trials. The support was primarily to reduce the financial risk of running numerous projects concurrently, and to accelerate the assessment process. The supported research project can still be terminated if the initial clinical results reveal serious safety issues or very low efficacy. For example, although MERCK received 38 million USD for COVID-19 vaccine research and development, the pharmaceutical giant still had to terminate its two vaccine research projects after the disappointing results of the Phase I clinical trials were revealed (MERCK, 2021). The review process helped protect people from undue harm that might be caused by ineffective vaccines.

5.3   Unfolding Vaccine Efficacy

However, I argue that the duty to protect cannot be used to justify the prioritisation of certain vaccines, where all candidates have been shown to be safe and effective. For example, initial vaccine efficacy results suggested that messenger RNA vaccines like the Pfizer-BioNTech vaccine and Moderna vaccines outperformed the AstraZeneca vaccine and the Johnson & Johnson vaccine by around 25%. However, a closer look at the design of these vaccines' clinical trials reveals that comparing the efficacy of different vaccines might not be as helpful as we hope (Ledford, 2021).

First, although the clinical trials shared a similar structure, they did not follow an identical design. Such discrepancies in trial design make a direct comparison of figures pointless. Take, for example, the Johnson & Johnson and Pfizer-BioNTech vaccines. At first glance, the Johnson & Johnson vaccine seems less effective than the Pfizer-BioNTech vaccine. Clinical trials showed that the Johnson & Johnson vaccine was only about 70% effective compared to the 95% effectiveness of the Pfizer-BioNTech vaccine. However, the two figures cannot be directly compared because the setup of the trials was different. In the case of the Johnson & Johnson vaccine, the stated efficacy was against symptomatic COVID-19 15 days after the first dose. As for the Pfizer-BioNTech vaccine, the 95% efficacy was about the effectiveness against symptomatic COVID-19 7 days after the second dose.

Second, the trials took place at different places and times. This is relevant in the context of a fast-evolving pandemic situation, as COVID-19's prevalence changed significantly in different places at different times. Conducting a clinical trial at a time and place with a relatively low prevalence of COVID-19 means that many participants might not be exposed to the virus at all. This can inflate the efficacy result. For example, the Pfizer-BioNTech and Moderna vaccines trials were conducted around the same time – when COVID-19 cases per capita were relatively low (around 20–40 cases per 100 k in the United States). However, when the Johnson & Johnson vaccine was trialled, Covid-19 cases per capita had grown to 40–80 cases per 100 k in the United States. Furthermore, most of the trials were conducted primarily in South Africa and Brazil, where the COVID-19 case rates were higher. The relatively higher prevalence might have impacted the results of Johnson & Johnson vaccine's efficacy against sympotematic COVID-19.

Third, the dominant variants presented in the clinical trials were also different. The more infectious Beta variant was identified in South Africa (where the Johnson & Johnson vaccine was being tested) shortly after the trial began. Something similar occurred in Brazil. After Johnson & Johnson's trial took place in late 2020, the more contagious Zeta variant quickly became the dominant variant in the country. These changes were reflected in the clinical trials. For example, 67% of the infected cases from Johnson & Johnson's trial in South Africa were the Beta variant. In contrast, most of the infections in the Pfizer-BioNTech trial were with the original, less infectious, variant.[3]

Due to these factors, clinical trial results are best understood as a snapshot of how effective the vaccine under study was at a particular time in a particular region. Had the Johnson & Johnson vaccine been tested earlier and against the original strain only, it may have demonstrated similar, or even better, effectiveness than the Phizer vaccine – or not. Effectiveness figures cannot, therefore, be meaningfully compared.

Furthermore, even if effectiveness could be meaningfully compared, prioritising certain vaccines over others at the expense of suspsending part of the vaccine programme can cause more harm than good if the goal of vaccination is not to eliminate COVID-19 but to reduce serious consequences of disease. In an interview with VOX, Dr Amesh Adalja at the Johns Hopkins University Center for Health Security pointed out that

> The goal of a vaccine programme for COVID-19 is not necessarily to get to 'COVID zero', but it's to tame this virus, to defang it, to remove its ability to cause serious disease, hospitalisation, and death. (Vox, 2021)

In other words, if we shift our focus to how effective a vaccine is at preventing severe symptoms and hospitialisations, then the data currently available to us shows that the Johnson & Johnson and the AstraZeneca vaccines are as good as the Pfizer-BioNTech and Moderna vaccines (de Gier et al., 2021).

5.4 Uncertainties, Risks, and Incommensurability

Theoretically speaking, the problems highlighted in Sect. 5.3 could be addressed by requiring all vaccine research teams to perform clinical trials simultaneously, with the same demographic makeup, at the same location. Once all of these factors are

---

[3] For a comparison between the time periods and the dominant variants presented in Pfizer-BioNTech's and Johnson & Johnson's clinical trials, see Vox, 2021.

controlled, it would then become possible to compare the efficacy of different vaccines and prioritise certain vaccines.

Indeed, we could improve protocols for conducting clinical trials during a pandemic. However, even if we could control these factors without delaying vaccine development, unknowns would remain. Take the Pfizer-BioNTech and Moderna vaccines as an example. Puranik et al. (2021) found that even though the two vaccines were based on the same technology (i.e. messenger RNA) and performed similarly in early trials, it is still challenging, if not impossible, to predict their efficacy against new variants. Puranik et al. observed that the Pfizer-BioNTech vaccine's efficacy against symptomatic COVID-19 dropped significantly to 42% six months after the research was initiated in January 2021 in the United States. While the Pfizer-BioNTech vaccine's efficacy declined significantly, the Moderna vaccine remained highly effective against symptomatic COVID-19 (76%). This information could not have been available when rollouts started.

New data gathered in the UK also shows that vaccines that provide better short-term protection do not necessarily outperform other vaccines in the long run. For example, Pouwels et al. (2021) found that the efficacy of the Pfizer-BioNTech vaccine dropped faster than that of the AstraZeneca vaccine. The trend suggests that after 20 weeks of inoculation with the second dose, the Pfizer-BioNTech vaccine becomes less effective than the AstraZeneca vaccine at providing protection against symptomatic COVID-19. Currently, scientists still don't know why Pfizer-BioNTech's efficacy declines so quickly (a 22% decline in 90 days).

Experts also anticipate that long-term safety issues may arise later. Previous research on an Ad5-based HIV vaccine found that the vaccine not only failed to protect against HIV, it actually increased the vaccine recipient's chances of contracting the virus. Some scientists warn that COVID-19 vaccines using similar technology, such as CanSino Biologics' Convidecia and Gamaleya's Sputnik V, might also increase the risk of contracting HIV in the long run (Kim et al., 2021). During the composition of this chapter, the European Medicines Agency is investigating the risk of developing a rare inflammatory condition called multisystem inflammatory syndrome from receiving the Pfizer-BioNTech vaccine and the risk of developing venous thromboembolism from receiving the Johnson and Johnson vaccine (Reuters, 2021). While out understanding of the vaccines continuously increase, it is still too early to tell whether there will be long-term safety issues.

It is also uncertain which vaccine will be the most effective against newer variants. For instance, a Canadian research team found that at 14 days after the first vaccine dose, the Pfizer-BioNTech vaccine was 2% more effective against the symptomatic COVID-19 of the Alpha variant than the AstraZeneca vaccine, but that the AstraZeneca vaccine was 12% more effective against the symptomatic COVID-19 of the Delta variant than the Pfizer-BioNTech vaccine (Nasreen et al.,

2021).[4] This research suggests that an initially successful vaccine might not outperform other vaccines in terms of its efficacy against all variants. Given that the COVID-19 is still mutating, rather than providing 'better alternatives', tyring to prioritise certain vaccines over others might be more akin to putting all the eggs into one basket.

The cases presented here show that attempts to prioritise certain vaccines over others cannot be epistemically justified. Options that seem superior may turn out to be inferior as our understanding of the vaccine increases and as the disease context changes. For instance, Israel decided to revise its exclusivly messenger RNA vaccine programme and add the adenovirus vector-based AstraZeneca vaccine to its vaccine pool in late 2021, even though this vaccine was considered 'inferior' by some states in early 2021 (Tercatin, 2021). Israel's response highlights that even when decisions are made following incomplete but best-available data, it is important that flexibility to revisit those decisions be maintained.

Yet, even if there is sufficient scientific evidence supporting the claim that a specific vaccine is better, this does not mean that a liberal state may thus prioritise the vaccine at the expense of suspending part of a vaccine rollout. It is frequently overlooked in the discussion of the 'better alternatives' argument that each available option is associated with various risks and benefits that might not be commensurable (Chang, 1997). Appealing to the duty to protect people from a certain risk at the expense of exposing that to a different set of risks provides little justification for the suspension and prioritization (Huang, 2021).

No matter which vaccine a person decides to take (or not take), they will have to bear the risk of unwanted side effects and, sometimes, symptoms that are not expected by medical experts. This is part of why the idea of a compulsory COVID-19 vaccination programme remains highly controversial. More rare but severe symptoms may emerge later in the future. Although this is thought to be unlikely, we cannot know for sure. Remaining unvaccinated also exposes one to a different set of risks. The first quarter of 2021 saw a resurgence of confirmed cases of COVID-19, with more than 10 million new cases reported to the World Health Organization (2021) in the first two weeks of April 2021.

The delay caused by vaccine rollout suspensions meant that many people could not take immediate and statistically effective action to reduce their risk of contracting COVID-19. From this perspective, the suspension or deliberate delay of a vaccine rollout forces people to bear risks they do not want to bear. The risks a person will have to take when they decide to undergo a vaccination are categorically

---

[4] This research was based on the data collected during December 2020 to May 2021 in Ontario, Canada. Many data points, such as the Moderna and AstraZeneca vaccines' effectiveness against symptomatic COVID-19 7 days after the second dose, were not presented in the research, likely because Canada only began its vaccination programme in December 2020 As a result, while the data used by this research indicated that the Pfizer-BioNTech vaccine performed less well than the AstraZeneca vaccine under certain circumstances, it is too early to draw a definitive conclusion. Nevertheless, my point holds: that an initially successful vaccine might not outperform other vaccines in terms of its efficacy against all variants. Indeed, as I write, there is a scramble to determine the effectiveness of various vaccines against the newly emerged Omicron variant, and similar issues will arise for future variants.

different to those one will need to bear when remaining unvaccinated. Hence, it is problematic if a person is only allowed to take the risks of remaining unvaccinated but not the risks associated with (presumed to be) less effective vaccines.

The fact that many countries still have not introduced compulsory measles vaccination despite overwhelming scientific proof of its efficacy and safety shows that sometimes vaccine efficacy and safety are not the only ethical consideration we need to take into account. Smoking presents a useful related example. There is substantial evidence that smoking increases the health risks of developing several severe diseases, such as lung cancer and coronary heart disease (National Health Service, 2018). It is estimated that smoking causes more than 480,000 deaths each year in the United States alone (US Department of Health and Human Services, 2014). However, most countries only regulate tobacco use in public spaces such as hospitals, schools, and libraries. Very few, if any, have introduced a categorical ban on tobacco.

The rationale behind the regulations is closely aligned to John Stuart Mill's (2003) Harm Principle. According to this Principle, the only occasion where a government can justifiably exercise its power over any member of society, against their will, is to prevent harm to unconsenting others. If a smoker is only to increase their own health risks, they are entitled to do so. Yet, smoking in public spaces might increase the health risks of others against their will. Therefore, it is justifiable for the state to restrict the smoker's freedom to smoke in public spaces.

The ethical foundation of vaccine prioritisation and the suspension becomes shaky once we compare this approach to vaccination with other health-related policies. So long as the risk of harm is limited to the decision-maker, the government should not intervene in a person's decision. Currently, COVID-19 continues to cause an enormous number of deaths each day. Taking away a person's opportunity to be vaccinated with a vaccine that is available and clinically shown to be safe and effective is to force them to remain exposed to the risks of contracting COVID-19. This damages the person's ability to act upon their decision and fails to pay due respect to their right to decide which risks they deem worth taking (Huang, 2021).

5.5 Duty to Facilitate Risk-Taking

One might argue that suggesting that there is a right to take risks is absurd because it implies a duty to facilitate risk-taking. A Millian liberal might concede that a liberal state has a negative duty not to interfere with risky behaviour so long as the behaviour does not directly negatively impact other people's. Yet, positively supporting risk-taking is another matter. If the right violated by certain liberal states were the right to take risks, then the way the states violate this particular right is by refusing to proactively provide their people with vaccines deemed to be inferior. Following this rationale, it seems that anyone interested in having a psychedelic experience or using hard drugs likewise has a right to demand the state facilitate their engagement with these substances.

Indeed, the duty to facilitate risk-taking might sound strange at first. Yet, the fact that most liberal states do not forbid their citizens from smoking or travelling to malaria-endemic regions suggests otherwise. Information printed on cigarette packages in some countries, like statements that smoking increases the risk of developing lung cancer, can be seen as a soft deterrent. However, such a message is also a piece of information aiming to help individuals decide whether the risk is worth taking. The same applies to anti-malaria drugs. Malaria is a severe infectious disease that can cause symptoms such as seizures and comas, and in some cases, death (Caraballo & King, 2014). There's no doubt that malaria poses a severe health threat to healthy individuals. Hence, it is understandable that many countries *advise against* unnecessary travel to malaria-endemic regions. But instead of dictating that no one should take the risk of contracting malaria, most liberal states *help* their citizens decide whether to take the risk, and how to mitigate the risk, by providing detailed travel information and anti-malaria drug information.

The duty to facilitate risk-taking is not a duty to help people take whatever risks they deem worth taking. The primary consideration here is to facilitate good decision making and to respect value pluralism. The reason a liberal state has a duty to provide malaria-relevant information to its people is not that exposing oneself to malaria is worth pursuing in and of itself, but that it is reasonable for one to value the experience of travelling to a malaria-endemic region.

The idea of reasonableness may help us distinguish between the cases of abusing hard drugs and receiving a less effective vaccine. The cases I presented in Sect. 5.4 show that even if we only consider relevant scientific facts, there is nevertheless much room for reasonable disagreement (Ismaili M'hamdi, 2021; Scanlon, 1998). For instance, many public health experts argue that reducing hospitalisation should be prioritised, whereas some politicians believe offering individual vaccine recipients better protection against COVID-19 is more critical. While the goals posited by the two views are very different, this does not mean that one of the two views must be wrong. Sometimes, differences in priority only show that people have different conceptions of the good and prioritise different values.

In the COVID-19 context, several considerations can be reasonably prioritised. One may prioritise convenience over efficacy and opt for the Johnson & Johnson vaccine (where it is readily available). One may prioritise gaining immunity as quickly as possible, opting for the first available vaccine that can provide sufficient protection. One may prioritise gaining immunity against COVID-19 over the concern of developing rare but severe symptoms like cerebral venous thrombosis (and be happy to take the AstraZeneca vaccine). Conversely, one may prioritise avoiding a vaccine with known but rare risks in favour of waiting for a vaccine that has fewer known risks, as did people who chose to avoid AstraZeneca and wait for other vaccines to become available to them. Likewise, people who decide to receive COVID-19 vaccination prioritise gaining immunity against COVID-19 over the risk of developing known rare short term complications, and over the possible risk of unknown health issues from vaccination. These prioritisations are all reasonable and open to disagreement.

Yet this is not to say that all disagreement is reasonable. Consider the concern that COVID-19 vaccines are not safe because they were developed and deployed very quickly relative to standard pharmaceutical development timelines. The concern is not entirely ill-founded. Given that most vaccine development takes more than a decade to enter the clinical trial phase (Hanney et al., 2020), it is understandable that some might think that the COVID-19 vaccine development must not have gone through all the necessary scrutiny. However, this concern can be easily clarified once one is adequately informed of the details of Operation Warp Speed (e.g. the financial support that allowed parallel research and development on multiple vaccine candidates and the administrative support that accelerated the review process of clinical trials).[5] Similarly, whether or not drinking bleach can prevent COVID-19 is not open to reasonable disagreement. It simply doesn't work.[6]

It is important to recognise that life is never risk-free. In the context of the COVID-19 pandemic, no matter which vaccine one eventually decides to take, one has to accept the risk of unwanted side effects, including the possibility of side effects unforeseen at the time of vaccination. This is another reason why compulsory COVID-19 vaccination remains highly controversial. Since we only have limited knowledge of COVID-19 and the available vaccines, implementing a compulsory programme will force people to take risks they might not be willing to take. From a right-to-take-risks angle, suspending part of a vaccine rollout to wait for a more preferred vaccine is equally problematic, as waiting for a different vaccine (or choosing to avoid vaccination) likewise carries risk. Currently, COVID-19 continues to cause an enormous number of deaths each day, with greater numbers of people facing severe illness and ongoing "Long COVID" symptoms. Depriving people of the opportunity to be vaccinated as soon as an effective vaccine is available forces them to continue to be exposed to the risks of contracting COVID-19.

---

[5] Financial constraints are part of the reason why vaccine developments usually take more than a decade. To reduce financial risk, a research team usually only works on one candidate at one time. Only after the team found that the candidate couldn't achieve the desirable results or meet the safety requirements, can the team move on the next candidate. Were it be possible to work on different candidates at the same time, it would not have taken so long for the research team to find the vaccine candidate that is both safe and effective (Hanney et al., 2020). Programmes like the Operation Warp Speed contributed significantly in terms of relieving vaccine developers of financial risk and made it possible for the developers to work on multiple vaccine candidates at the same time. Without financial support, MERCK probably would not have been able to afford to take the risk of starting two vaccine research projects at the same time. However, such risk-taking was important to ensuring that safe and effective vaccines would be found quickly. For more information on the Operation Warp Speed, see Slaoui and Hepburn (2020).

[6] There is much dangerous misinformation circulating on the internet. One example was the claim, debunked by the French government, that snorting cocaine helps protect people from contracting COVID-19 because the snorting can sterilize one's nostrils (Gregory, 2020). Chemical substances like methylene chloride and chloride dioxide were also falsely marketed as COVID-19 disinfectants (Dlouhy, 2020).

5.6   Fostering Trust by Facilitating Risk-Taking

Another reason for taking the duty to facilitate risk-taking seriously in times of uncertainty is to foster trust. While our knowledge of the COVID-19 virus and the short-term efficacy of different vaccines against different variants continues to grow, there are still many unknowns. It is hard to predict if there will be new variants that are more infectious or more deadly. In addition, the long-term efficacy of different vaccines can only be revealed with time. These uncertainties need to be appropriately communicated.

Regrettably, most liberal states failed to communicate the knowns and the unknowns to their citizens appropriately. The desire to increase vaccine coverage as quickly as possible led many states to focus on conveying messages regarding the effectiveness and safety of the vaccines, while obscuring the admittedly small health risks associated with vaccination. Understandably, some people became hesitant after they learned about cerebral venous thrombosis. However, the vaccine rollout suspensions didn't offer any meaningful clarification, they simply added to the confusion. It's not surprising that after the decision to suspend the AstraZeneca vaccine's use, vaccine hesitancy rose in European countries by 9% (Ahrendt et al., 2021; Ellyatt, 2021). The suspensions 'confirmed' people's suspicions that vaccines were not as safe as the states had claimed, and that there might be information not properly revealed to the general public.

The issue here is that, while states may not have set out to overpromise on vaccines, the optimistic tones they adopted makes it appear as if they did. The failure to properly address people's concerns further weakened already fragile trust – if a vaccine that was promoted as safe and effective turned out to be not as safe and effective as promised, this left open the possibility that other vaccines might likewise be less safe than currently claimed. This distrust could have been mitigate by acknowledging that while the clinical trials were conducted in a very rigorous manner, there remained a possibility of rare but severe symptoms showing up after the commencement of large-scale vaccine rollouts. Take the risk of developing cerebral venous thrombosis as an example. A liberal state could help its people decide whether it is worth taking the risk of developing cerebral venous thrombosis from receiving an AstraZeneca vaccine by providing the information that the risk of developing cerebral venous thrombosis from COVID-19 is roughly eight-times higher than from receiving the vaccine (Taquet et al., 2021).

5.7   Conclusion

In this chapter, I developed a duty-based critique of COVID-19 vaccination policies. This is not to disregard the importance of risk-benefit analysis. Fighting against a public health crisis like the COVID-19 pandemic requires input from the latest epidemiological data and careful analysis of the risks and benefits of each available option. However, given epistemic limitations and the incommensurability of different risks and benefits, a consequentialist risk-benefit framework is not always helpful. In situations of uncertainty, a duty-based framework may offer more stable normative guidance that will not be easily undermined by constantly changing epidemiological data. Devising counter-Covid-19 strategies based on this approach upholds vital liberal principles and reduces the likelihood of creating confusion for the general public.

A liberal state does have a general duty to promote people's well-being and safeguard its people's lives from undue health risks. However, as we are currently in a situation where no one knows which vaccine will be the most effective against newer variants, will have the fewest long-term side effects, or will provide the longest-lasting protection, it is doubtful that a liberal state may legitimately decide which of the available options is *best* on its people's behalf. Moreover, even if these uncertainties are clarified, it is still morally unacceptable for a liberal state to prioritise certain vaccines at the expense of suspending part of the vaccine rollout.

A liberal state should acknowledge uncertainties, communicate to the public the known risks and benefits of each currently available option, and assist the public in taking what risks they deem best for their well-being.[7]

References

Ahrendt, D., Mascherini, M., Nivakoski, S., & Sándor, E. (2021). *Living, working and covid-19.* Retrieved from https://www.eurofound.europa.eu/publications/report/2020/living-working-and-covid-19 https://doi.org/10.2806/76802.

Baden, L. R., El Sahly, H. M., Essink, B., Kotloff, K., Frey, S., Novak, R., ... Group, C. S. (2021). Efficacy and safety of the mrna-1273 sars-cov-2 vaccine. *New England Journal of Medicine, 384*(5), 403–416. https://doi.org/10.1056/NEJMoa2035389

Caraballo, H., & King, K. (2014). Emergency department management of mosquito-borne illness: Malaria, dengue, and west nile virus. *Emergency Medicine Practice, 16*(5), 1–23.

Chang, R. (1997). *Incommensurability, incomparability and practical reason.* Harvard University Press.

Daniels, N. (2017). *Justice and access to health care.* The Stanford Encyclopedia of Philosophy. Retrieved from https://plato.stanford.edu/archives/win2017/entries/justice-healthcareaccess/

Danish Health Authority. (2021). *Denmark continues its vaccine rollout without the covid-19 vaccine from astrazeneca.* Retrieved from https://www.sst.dk/en/English/news/2021/Denmark-continues-its-vaccine-rollout-without-the-COVID-19-vaccine-from-AstraZeneca

de Gier, B., Kooijman, M., Kemmeren, J., de Keizer, N., Dongelmans, D., van Iersel, S. C. J. L., ... van den Hof, S. (2021). Covid-19 vaccine effectiveness against hospitalizations and icu admissions in the Netherlands, april–august 2021. *medRxiv.* https://doi.org/10.1101/2021.09.15.21263613

[7] I would like to thank Hafez Ismaili M'hamdi and Jilles Smids for pushing me to investigate the idea of the duty to facilitate risk-taking during the research seminar at the Erasmus Medical Centre. I would also like to give special thanks to Lucy Valenta for giving me invaluable adviace and helping me proofread the manuscript. I'd also like to thank the participants of the OZSW Annual Conference 2021 and the anonymous reviewers for giving me very constructive feedback.

Dlouhy, J. (2020, June 11). Epa tells amazon, ebay to stop shipping unproven covid goods. *Bloomberg*. Retrieved from https://news.bloomberglaw.com/coronavirus/epa-orders-amazon-ebay-to-stop-shipping-unproven-covid-products

Ellyatt, H. (2021). The damage is done': Europe's caution over astrazeneca vaccine could have far-reaching consequences. *CNBC*. Retrieved from https://www.cnbc.com/2021/03/16/europes-suspension-of-astrazenecas-covid-vaccine-is-damaging.html

Gregory, A. (2020). France tells citizens cocaine cannot protect against coronavirus. *The Independent*. Retrieved from https://www.independent.co.uk/news/world/europe/coronavirus-cocaine-france-government-warning-conspiracy-theories-disinformation-smurfs-a9389146.html

Hanney, S. R., Wooding, S., Sussex, J., & Grant, J. (2020). From covid-19 research to vaccine application: Why might it take 17 months not 17 years and what are the wider lessons? *Health Research Policy Systems, 18*(1), 61. https://doi.org/10.1186/s12961-020-00571-3

Huang, P.-H. (2021). Covid-19 vaccination and the right to take risks. *Journal of Medical Ethics*. https://doi.org/10.1136/medethics-2021-107545

Ismaili M'hamdi, H. (2021). Neutrality and perfectionism in public health. *American Journal of Bioethics, 21*(9), 31–42. https://doi.org/10.1080/15265161.2021.1907479

Katella, K. (2021). *Comparing the covid-19 vaccines: How are they different?* Retrieved from https://www.yalemedicine.org/news/covid-19-vaccine-comparison

Ketella, K. (2021). Comparing the covid-19 vaccines: How are they different? *Yale Medicine*. Retrieved from https://www.yalemedicine.org/news/covid-19-vaccine-comparison

Kim, J. H., Hotez, P., Batista, C., Ergonul, O., Figueroa, J. P., Gilbert, S., ... Bottazzi, M. E. (2021). Operation warp speed: Implications for global vaccine security. *The Lancet Global Health, 9*(7), e1017–e1021. https://doi.org/10.1016/s2214-109x(21)00140-6

Le, T. T., Cramer, J. P., Chen, R., & Mayhew, S. (2020). Evolution of the covid-19 vaccine development landscape. *Nature Reviews Drug Discovery, 19*(10), 667–668. https://doi.org/10.1038/d41573-020-00151-8

Ledford, H. (2021). Why covid vaccines are so difficult to compare. *Nature, 591*(7848), 16–17.

MERCK. (2021). *Merck discontinues development of sars-cov- 2/covid-19 vaccine candidates; continues development of two investigational therapeutic candidates* [Press release].

Mill, J. S. (2003). In M. Warnock (Ed.), *Utilitarianism and on liberty: Including mill's 'essay on bentham' and selections from the writings of jeremy bentham and john austin* (Second ed.). Blackwell Publishing.

Nasreen, S., He, S., Chung, H., Brown, K. A., Gubbay, J. B., Buchan, S. A., ... Kwong, J. C. (2021). Effectiveness of covid-19 vaccines against variants of concern, Canada. *Medrxiv*. https://doi.org/10.1101/2021.06.28.21259420

National Health Service. (2018). *What are the health risks of smoking?* Retrieved from https://www.nhs.uk/common-health-questions/lifestyle/what-are-the-health-risks-of-smoking/

National Health Service. (2021). *Storage requirements for each covid-19 vaccine*. Retrieved from https://www.sps.nhs.uk/articles/storage-requirements-for-each-covid-19-vaccine/

Polack, F. P., Thomas, S. J., Kitchin, N., Absalon, J., Gurtman, A., Lockhart, S., ... Group, C. C. T. (2020). Safety and efficacy of the bnt162b2 mrna covid-19 vaccine. *New England Journal of Medicine, 383*(27), 2603–2615. https://doi.org/10.1056/NEJMoa2034577

Pouwels, K. B., Pritchard, E., Matthews, P. C., Stoesser, N., Eyre, D. W., Vihta, K. D., ... Walker, A. S. (2021). Effect of delta variant on viral burden and vaccine effectiveness against new sars-cov-2 infections in the UK. *Nature Medicine, 27*(12), 2127–2135. https://doi.org/10.1038/s41591-021-01548-7

Puranik, A., Lenehan, P. J., Silvert, E., Niesen, M. J. M., Corchado-Garcia, J., O'Horo, J. C., ... Soundararajan, V. (2021). Comparison of two highly-effective mrna vaccines for covid-19 during periods of alpha and delta variant prevalence. *medRxiv*. https://doi.org/10.1101/2021.08.06.21261707

Remmel, A. (2021). Why is it so hard to investigate the rare side effects of covid vaccines? *Nature*. https://doi.org/10.1038/d41586-021-00880-9

Reuters. (2021). *Eu reviewing risk of rare inflammation after covid-19 vaccinations.* Retrieved from https://www.reuters.com/business/healthcare-pharmaceuticals/eu-reviewing-cases-rare-body-inflammation-after-covid-19-shots-2021-09-03/

Sadoff, J., Le Gars, M., Shukarev, G., Heerwegh, D., Truyers, C., de Groot, A. M., ... Schuitemaker, H. (2021). Interim results of a phase 1-2a trial of ad26.Cov2.S covid-19 vaccine. *New England Jolurnal of Medicine, 384*(19), 1824–1835. https://doi.org/10.1056/NEJMoa2034201

Scanlon, T. (1998). *What we owe to each other.* Harvard University Press.

Slaoui, M., & Hepburn, M. (2020). Developing sage and effective covid vaccines – Operation warp speed's strategy and approach. *New England Journal of Medicine, 383*(18), 1701–1703.

Taquet, M., Husain, M., Geddes, J. R., Luciano, S., & Harrison, P. J. (2021). Cerebral venous thrombosis: A retrospective cohort study of 513,284 confirmed covid-19 cases and a comparison with 489,871 people receiving a covid-19 mrna vaccine. *EClinicalMedicine, 39*, 101061. https://doi.org/10.1016/j.eclinm.2021.101061

Tercatin, R. (2021). *Covid: Israel to give astrazeneca vaccine to those affected by pfizer.* Retrieved from https://www.jpost.com/health-and-wellness/coronavirus/covid-israel-to-give-astrazeneca-vaccine-to-those-affected-by-pfizer-682346

Torjesen, I. (2021). Covid-19: Omicron may be more transmissible than other variants and partly resistant to existing vaccines, scientists fear. *BMJ, 375*, n2943. https://doi.org/10.1136/bmj.n2943

United Nations. (1948). Universal declaration of human rights. *UN General Assembly, 302*(2), 14–25.

US Department of Health and Human Services. (2014). *The health consequences of smoking—50 years of progress: A report of the surgeon general.* US Department of Health and Human Services.

van Dongen, A., & van Mersbergen, C. (2021, April 24). Rivm-vaccinatiebaas: 'Miljoenen doses astrazeneca die na half mei komen, zijn niet meer nodig'. *Algemeen Dagblad.* Retrieved from https://www.ad.nl/binnenland/rivm-vaccinatiebaas-miljoenen-doses-astrazeneca-die-na-half-mei-komen-zijn-niet-meer-nodig~a03948a0/

Vox (Producer). (2021). *Why you can't compare covid-19 vaccines.* Retrieved from https://www.youtube.com/watch?v=K3odScka55A

Voysey, M., Clemens, S. A. C., Madhi, S. A., Weckx, L. Y., Folegatti, P. M., Aley, P. K., ... Zuidewind, P. (2021). Safety and efficacy of the chadox1 ncov-19 vaccine (azd1222) against sars-cov-2: An interim analysis of four randomised controlled trials in Brazil, South Africa, and the UK. *The Lancet, 397*(10269), 99–111. https://doi.org/10.1016/s0140-6736(20)32661-1

World Health Organization. (2021). *Who coronavirus (covid-19) dashboard.* Retrieved from https://covid19.who.int/

Open Access This chapter is licensed under the terms of the Creative Commons Attribution 4.0 International License (http://creativecommons.org/licenses/by/4.0/), which permits use, sharing, adaptation, distribution and reproduction in any medium or format, as long as you give appropriate credit to the original author(s) and the source, provide a link to the Creative Commons license and indicate if changes were made.

The images or other third party material in this chapter are included in the chapter's Creative Commons license, unless indicated otherwise in a credit line to the material. If material is not included in the chapter's Creative Commons license and your intended use is not permitted by statutory regulation or exceeds the permitted use, you will need to obtain permission directly from the copyright holder.

Chapter 6
Conspiracism as a Litmus Test for Responsible Innovation

Eugen Octav Popa and Vincent Blok

6.1 The Edges of Inclusion

The inclusion of stakeholders in scientific and technological decision-making lies at the heart of many contemporary approaches captured under the umbrella term of *responsible innovation*. For example, inclusion is central to the 'standard' responsible innovation approach (Grunwald, 2011; Owen et al., 2013; van den Hoven et al., 2014), to technology assessment in its various versions (Grunwald, 2009; Hellstrom, 2003), ethics of technology (Groves, 2009; Hansson, 2017), and very clearly in the field of public engagement with science (Selin et al., 2017; Stilgoe et al., 2014; Wilsdon & Willis, 2004). While these approaches are different from each other in various ways, there is a strong consensus around the idea that non-scientific stakeholders should be included in innovation. Moreover, there are pragmatic reasons for it, since innovation thus becomes better and more sustainable, and moral ones, since innovation emerges from a democratic process of participation.

Are conspiracists part of this new deal? By 'conspiracists', we mean simply someone who believes or defends a conspiracy theory regarding a specific public product – in our case, a scientific theory or technological product. The conspiracist claims that the event in question results from hidden forces that conspire to pursue

E. O. Popa (✉)
Department of Science Technology and Policy Studies University of Twente,
Enschede, The Netherlands
e-mail: o.popa@utwente.nl

V. Blok
Department of Communication, Philosophy and Technology, Wageningen University
and Research, Wageningen, The Netherlands
e-mail: Vincent.Blok@wur.nl

© The Author(s) 2022
M. J. Dennis et al. (eds.), *Values for a Post-Pandemic Future*, Philosophy
of Engineering and Technology 40, https://doi.org/10.1007/978-3-031-08424-9_6

their (usually malevolent) interests.[1] Of the many questions that the COVID-19 pandemic has brought to the fore, one of the most relevant one for the field of responsible innovation is whether conspiracists are part of this new inclusive deal. If they are, it is necessary to understand how to approach the conflict between conspiracism and science. But if conspiracists are not part of this new deal, then we must ask under what conditions conspiracists have indeed forfeited their right to inclusion. In either case, all those who have sung the anthem of inclusion will probably see conspiracism as a challenge and perhaps a litmus test of how far the new deal can be extended.

The topic of conflict between various publics and science is not foreign in the field of responsible innovation. In the process of engaging stakeholders with different institutional and psychological profiles in science, 'friction' and competition between incompatible perspectives (*agonism*) is bound to occur (Popa et al., 2020b). Acknowledging this, some scholars have pressed the point that conflict must be not avoided, but in fact, sought and encouraged for its practical and moral benefits (Jasanoff, 2003; Cuppen, 2012; Timmermans & Blok, 2018). But even scholars that are generally welcoming of inter-stakeholder conflict tend to restrict their view to standard conflicts that are manageable in principle, and that can be traced back to a discernible difference in the knowledge that the parties have or the values that they accept (for a discussion of this limitation, see Blok, 2019). Such standard conflicts can undoubtedly be satisfactorily managed with our existing tools for participation: stakeholder workshops, consensus conferences, panels, focus groups and the like.

By contrast, conspiracists reside at least *prima facie* at the edges of and perhaps even outside the above-described concept of inclusion. Their opposition to mainstream science institutions – either in general or concerning a specific scientific output – is much more definitive than that of the typical sceptic (Byford, 2011; Coady, 2006). For conspiracists, the game is not played to their disadvantage; the game is altogether rigged. In a post-truth society of 'alternative facts', the conspiracist position is increasingly being taken and thus increasingly normalised (Fuller, 2018). So how can the game of responsible research and innovation be played with those who believe that the game of research and innovation is rigged? Understanding the relationship between conspiracism and responsible innovation is necessary in order to understand the unvisited corners of the science-society interface in the post-pandemic future.

In this paper, we argue that pluralism can provide insights on how to tackle these questions. We maintain that conspiracists *qua conspiracists* have not forfeited their right to inclusion, first because the occurrence of conspiracies is not a logical impossibility – every so often conspiracies do occur – and second because they can, generally speaking, share common values with scientists as well as the rest of society. They might not always share the scientist's method of truth-finding and truth-testing, but by and large they are driven by worries that we can recognize, e.g., power

---

[1] This definition is generally in line with literature on conspiracism, but scholars sometimes need to make more precise terminological distinctions between sub-groups defined by specific aims or beliefs (Byford, 2011; Coady, 2006; Dentith, 2014, 2018).

monopolies, media not providing a faithful representation of reality, inequality between different social groups, health risks. If conspiracists are indeed part of this new deal, as we think they are, then pluralism can therefore provide insights to foster this inclusion. Specifically, we want to argue that pluralism is a beneficial philosophical starting point from which to strike a balance between two possible monist responses to conspiracists: *over-inclusion* of conspiracists (in which conspiracists are included by ignoring their radical conflict with, and distrust of, science) and *over-exclusion* of conspiracists (meaning that conspiracists are excluded as if their right to inclusion has been forfeited). In both cases, the monist response is an oversimplification. To illustrate what it means to approach the relationship between science and conspiracism from a monist perspective and to describe what it means to work towards a pluralist perspective, we discuss a real-life example of a *monist* response to conspiracism – the 2021 European Commission policy against conspiracism – and we explain what a *pluralist* alternative would look like.

In Sect. 6.2, we provide a general description of conspiracism. We focus on the conspiracist phenomenon as it occurred concerning the COVID-19 pandemic (a topic that will serve as a case in point for comparing monist and pluralist responses in Sect. 6.4). In Sect. 6.3, we outline pluralist philosophy to elucidate what it means to put forward a pluralist response to conflict. In Sect. 6.4, we take as a case in point a policy that illustrates the *monist* approach – the E.U. Commission guidelines for dealing with conspiracists in the case of the COVID-19 pandemic. In Sect. 6.5, we bring pluralist insights to bear on this case and show how the field of responsible innovation can mount a pluralist response to conspiracism. Finally, in Sect. 6.6, we compare pluralism and monism and discuss further challenges.

6.2   Conspiracism and Its Recent Occurrence During COVID-19 Pandemic

As mentioned above, we take the term 'conspiracism' in its broadest sense, including individuals who advance and argue for conspiracy theories (i.e., 'conspiracy theorists') and those who simply believe in the theories advanced by others (for general overviews and philosophical perspectives, see Coady, 2006; Dentith, 2018). But what is a conspiracy theory?

At first sight, the term 'conspiracy theory' requires little explanation: a conspiracy theory is a theory about a conspiring group of individuals. In this definition, we typically allow the term 'theory' to cut both ways: conspiracy theories are *theories₁* in that their epistemic function is to *explain* an event (or events), but they are also *theories₂* in the sense that they are not established facts but 'mere theory', that is, hypotheses (Coady, 2006; Dentith, 2018). Despite the concept of 'conspiracy theory' posing no particular semantic difficulties, it has been repeatedly pointed out that identifying conspiracy theories in real-life situations is hardly a straightforward task (Dentith, 2014, 2018). For example, not all groups with morally questionable intentions form a conspiracy and not all hypotheses about large-scale deceit count as conspiracy theories. It is not within the scope of this paper to work out these problems of denotation. Still, it is crucial to draw attention to them in order to understand the scope of the claims we will advance.

There is, however, one feature of conspiracy theories that has been recognised by nearly all who have tackled the subject in the past: conspiracists *are sometimes right*. This is particularly true in science, where the harmful effects of many products were uncovered after many years and with them the realisation that those who stood to profit from those products were very much aware of those harmful effects. Classic examples are asbestos, leaded gasoline, halocarbons, diethylstilboestrol (DES) and Tributyltin (TBT) (Gee, 2001; Harremoës et al., 2013). Admittedly, these cases might not fit the cliché picture of a macabre conspiracy for world domination. Still, they fit the definition of a small group of people with (mainly) economic interests, suppressing or ignoring evidence against the broader public. So, for example, if you had been a conspiracist in the 1960s claiming that the big oil companies such as Standard Oil and General Motors are suppressing evidence regarding the damaging effects of leaded gasoline and that the research confirming the safety of leaded gasoline was faulty and muddied by the interest of those companies if you had claimed that the companies were aware of alternative gasoline additives that were safer but less profitable yet decided to invest in (tetraethyl)lead instead causing long-term environmental problems that continue to the present day, you would have been correct, however improbable these claims may have sounded then (Needleman & Gee, 2013).

For the present purposes, all this is relevant because it constitutes the *prima-facie* case for the claim that conspiracists are part of the new inclusive deal. Conspiracists do not fall qua conspiracists outside the realm delineated by the core values we share (truth, fairness, safety etc.), conspiracists are sometimes right, and this seems to be a sufficient reason for including conspiracists in the ideal of 'science *with and for society*' (Owen et al., 2013). At the same time, we cannot forget that that conspiracy theorists can endanger the deployment of beneficial policies and thus lead to hazards and even the loss of human lives (Naeem et al., 2021; Romer & Jamieson, 2020). A case in point of such risks is the recent COVID-19 conspiracism.

By 'COVID-19' conspiracism, we mean the variety of conspiracy theories propounded between December 2019 and August 2021 around the COVID-19 pandemic. In March 2020, as the COVID-19 pandemic was advancing without any solution or vaccine in sight, the hashtag #FilmYourHospital was used on Twitter for user footage of hospitals or testing locations that were deemed too empty or too calm for a global pandemic. This footage, some users claimed, showed that the COVID-19 pandemic was either a hoax, meaning that the governments and scientists were lying about its existence, or, at the very least, that the pandemic was much less acute than the mainstream media and science had us believe. In one of these videos, a man is filming a relatively inactive testing location and repeatedly asks: "Where are all the sick people?". A Twitter account, now banned, sparked a debate in early February 2020 with the following text:

The survival rate of Coronavirus is nearly 98%. When you count young, healthy adults, it is closer to 99.5%. Why is this being marketed as The Black Plague? Democrats get to crash the economy, and the Chinese get protesters off the streets of Hong Kong (quoted in Pummerer et al., 2020)

All this happened in the beginning months of the COVID-19 pandemic. But as the pandemic progressed, actions of this kind continued and expanded in complexity. Just about every aspect of the pandemic has sprung some form of conspiracist thinking from the more garden-variety allegations of inflated death counts to the surprisingly resilient theory that the virus was released intentionally to secure Chinese economic supremacy or, the other way around, that the CIA created the virus to 'keep China down'.

Indeed, conspiracy theories around epidemics and pandemics are not a novel phenomenon, but it seems that the COVID-19 conspiracism was exacerbated by the widespread use of social media (Cinelli et al., 2020; Douglas et al., 2019; Enders et al., 2021; Ferrara et al., 2020; Larson, 2018; Naeem et al., 2021). The term 'infodemic' has been coined to draw a parallel between the spread of the COVID-19 virus, or viruses in general, and the spread of misinformation primarily through social media. Thus, not only is COVID-19 conspiracism more *present* in the public arena compared to past outbreaks of, say, AIDS, SARS and H1N1 (Lee, 2014), but its effect on actual policy and governmental responses is amplified (Naeem et al., 2021). Present-day conspiracism is just like its relatives from the past, but it is, to put it simply, both bigger and stronger. As a result, institutional *responses to* COVID-19 have been correspondingly more visible and more drastic in their rejection of whatever the conspiracists were claiming (see also below Sect. 6.4). Nevertheless, in what follows, we want to argue that conspiracists can be included in research and innovation – as we prima facie established they have a right to – but that this process requires the adoption of a pluralist philosophical stance. Therefore, we must introduce the notion of pluralism and explore the contrast between pluralism and monism in responding to conspiracism.

6.3   Pluralism and Monism

We claim that you must be a pluralist if you want to do justice to the conspiracist's *prima facie* right to inclusion. But what does it mean to be a pluralist? Pluralism can best be described as the philosophy that stands halfway between rationalism and relativism (Crowder, 2021, pp. 218–220). If rationalism is the conception that Reason must guide our answers to life's essential questions and relativism the notion that nothing is subject to such an ideal, pluralism is the conception that there are many ideals of Reason and that these multiple ideals can be incompatible with one another. The pluralist believes, as the rationalist does, that some choices are better than others – because they conform to an ideal or not –, but they also believe, as the relativist does, that there is no common, fundamental overarching ideal (Lassman, 2011). Although pluralism stands philosophically between rationalism and relativism, pluralists are constitutionally closer to relativism than rationalism. The historical roots of pluralism can be found in the relativism of Protagoras and the Skeptic school, yet pluralists do not deny the possibility of rational choices (Berlin, 1998; Kekes, 1993).

In contemporary philosophy, it was the work of Isaiah Berlin that has reignited the discussion on pluralism. Berlin advanced pluralism as an alternative to the reductionism characteristic of both those who always seek to follow Reason and those who always seek to undermine it. Later commentators have noted – and Berlin eventually confirmed – that Max Weber was a precursor in many of these ideas. Nowadays, scholars herald Weber as the first contemporary pluralist in Western philosophy that has worked out the consequences of a plural conception of the good (for overviews of this historical development, see Crowder, 2021; Lassman, 2011).

As one might expect, there is a plurality of versions of pluralism. These were developed primarily in political philosophy (Crowder, 2021; Hampshire, 2018; Kekes, 1993; Oakeshott, 1991; Walzer, 1983). More recently, the ideas developed in political philosophy have been applied to not only policymaking at various levels (Galston, 2005; Paxton, 2020) but to the study of the interface between science and society (Popa et al., 2020a; Stirling, 2008). The scope of this chapter doesn't need to elucidate all these evolving pathways. Instead, a general overview of several key ideas will be more helpful.

Basic Values

Pluralists start by acknowledging that there is a set of fundamental (or 'primary') values that are valid universally and are shared at a fundamental level between all members of society to leave little or no room for deviation (Crowder, 2021, pp. 118–122; Kekes, 1993, pp. 118–120). Plurality is thus said to be *restrained* by the idea that all reasonable human beings will agree on a baseline of human decency, meaning that "the protection of life, physical security, and some freedom to do as we please are normally good in all historical and cultural contexts" (Kekes, 1993, p. 119). The values are thus basic, yet they are not absolute. They are not absolute because: (i) some might not hold in extraordinary circumstances; (ii) we might choose one over the other in case they conflict; and (iii) there will be a diversity of interpretations of them across cultures and time periods. These three points, however, do not "extend so far as to call into question the truisms embodied in deep conventions that all conceptions of a good life require the protection of life, physical security, and some freedom from undeserved violations" (Kekes, 1993, p. 120)

There are thus various conceptions of the good life, but there is also something that all conceptions of the good life share, namely basic values. Pluralists do not deem it necessary to offer a complete description of this common core. (After all, the statement that 'There exists a common core' can be logically confirmed by even one singular example). For instance, according to Hampshire, there is a unifying conception of fairness shared by all cultures, which is formed around the principle of *audi alteram partem*, meaning that when a conflict arises, all involved parties must be allowed to state their case (Hampshire, 2018). Others point towards the shared value of life, which all traditions must protect one way or another, or some degree of individual liberty to live and do as one pleases. These are examples of fundamental values that constitute the core of social cohesion and thus a starting point for dealing with conflict.

Beyond Basic Values: Agonism and Contingency

Conflict is thus an inherent part of our social and political life, if only because of the mentioned incompatibility and incommensurability between 'basic values': the value of life will at times oppose the value of justice, the value of liberty will at times oppose the value of equality etc. The pervasiveness of this conflict is explained by the fact that many different conceptions of the good life satisfy various subsets of basic values. The most pressing question for a pluralist is understanding and criticising human behaviour under such conditions of plurality.

First, pluralists point out that rationalist dichotomies (truth/falsity, right/wrong, knowledge/opinion) constitute one facet of the social relationships developed between the parties involved in the conflict (Crowder, 2021, p. 75). It might seem like a truism to say that social relationships stand or fall based on something more than who's right and who's wrong. Yet, by claiming this, pluralists reject the age-old rationalist idea that that conflict must be approached rationally, neutrally, from what was referred to as the 'view from nowhere' (Nagel, 1986). It equally means that conflict must *not* be regarded as a Socratic, dialectical battle of the minds where speakers put forward rational argumentation, and the cases are being judged 'on the merits'. Instead, pluralists look for alternatives to this *cognitive* orientation towards conflict, alternatives that can justify the idea that there are many versions of the good life (Blok, 2019).

More important than ending up on the wrong side of truth is the risk of destroying or defusing the *agonism* between the parties involved (Blok, 2019; Mouffe, 2005; Paxton, 2020). Agonism refers to the adversarial relationship between individuals (or groups) who see each other as opponents but acknowledge their legitimacy in participating in opposition. Some prototypical examples: boxers in a ring are in a state of agonism; political parties in a parliament are, *de jure*, in a state of agonism. But agonism can 'decay' into a state of *antagonism* where the parties see each other not as adversaries but as enemies, meaning that one party allegedly defends the home base legitimately. In contrast, the other is the illegitimate intruder (in our case, mainstream Science defends truth legitimately while conspiracists are the illegitimate purveyors of fake news). The antagonistic relationship is characterised by division and fundamental distrust, usually maintained on both sides. Unlike adversaries, enemies are not just *different* claimants at the throne of reigning consensus; they constitute *the other* who needs to be eliminated from the prevailing consensus. It should be mentioned that pluralists recognise that antagonism is "ineradicable", meaning there will always be insiders and outsiders relative to a particular symbolic line surrounding groups, institutions, societies, nations etc. Our fundamental political duty is to find democratic orders through agonism despite this underlying remnant of antagonism (Mouffe, 2005).

Responding to Conflict: Monism Versus Pluralism

The philosophy of pluralism has direct implications for our approach to the conflict between conspiracists and mainstream science. Monism prescribes that those engaged in conflict – both representing the reigning consensus and contesters – must follow the rules derived from this one overarching good. In our case, people engaged in a conflict about a conspiracy theory must follow the rule of Reason, and they must do so without deviant ("irrational") discussion tricks or fallacies (Hansen & Pinto, 1995). Pluralists appreciate this but point out that being reasonable is not just a matter of avoiding fallacies but also resisting the temptation to view the conflict as "a crisis produced by our adversary's stupidity, wickedness, or perversity" (Kekes, 1993, p. 24). Furthermore, avoiding fallacies is not the same as recognizing the other as a legitimate adversary, that the parties seek the origin of the conflict, that they respect each other's identity etc. Being reasonable in this monist sense applies only to the parties' behaviour *within* the discussion whereas, as explained above, the pluralist urges us to look 'above' the discussion to the origin and the effects of that discussion on the parties' relationship.

> By taking the larger view suggested by conceptions of a good life or by traditions, we come to see the conflict at hand in a different light. We shall not merely ask: what should we do here and now? We shall ask instead: what should we do here and now so that we could resolve this conflict in a way that would be best from the point of view of the system of values we, as disputants, share? And if we are reasonable, we shall answer by stepping back from the immediacy of the conflict in which we participate in order to reflect on what would be best not here and now but in the long run, given the values of our tradition or our conception of a good life (Kekes, 1993, p. 25).

A pluralist approach to conflict means that the monist concept of a reasonable discussion (alignment with reason) becomes just one of the many ideals animating the discussion. It does so to maintain agonism and with an eye for the basic values shared by the parties involved.

The reader will have gathered from the above that a pluralist approach to conflict is not a ready-made method for conflict *resolution*. To be a pluralist about conflict is not to follow a particular method towards resolution, nor does it involve any particular longing for resolution. Instead, it focuses on responsiveness towards others – particularly those who have a worldview that differs from yours radically (Blok, 2019). It also involves interpreting what is happening – conflict – in a different way, not as a clash between someone who is right and someone who is wrong, or between someone who is informed and the other who is misinformed, but rather as a dialogue between two different, and possibly incompatible, identities (Kekes, 1993). Instead of 'fact checking' and 'fallacy finding' and 'debunking', the pluralist looks at the other as formulating an ethical demand in such a way that a response does not annul any of the identities engaged in the dialogue.

6.4 Monism in Policy Responses to COVID-19 Conspiracism

The European Commission and national governments within the European Union,[2] the World Health Organization[3], and social media organisations such as Twitter[4] and Facebook[5] have responded to the identified threat of COVID-19 conspiracism in remarkably similar ways. Yet, despite their difference in origin and field of application, what unites these policies is their *monism*. What does this mean? For illustration, we will take a closer look at the policy advised by the European Commission concerning conspiracism. The monist features illustrated here are the following:

(i)   A rationalistic framing of the discussion, focusing on one dominating value
(ii) A binary representation of truth (true vs false)
(iii) Discarding opponents as not just wrong but morally wrong
(iv) Formal invocation of pluralist toleration

A first feature is that the policy is cast in a rationalistic mould of truth-finding, fact-checking, debunking, fallacy identification, evidence testing, refutation and the like. Truth is the name of the game, and conspiracists are the ones losing. As examples of a discourse of rationalistic orientation, consider the italicised passages in the quotes below:

> Conspiracies [...] require to strengthen the commitment of the whole society, including competent authorities, media, journalists, fact-checkers, civil society and online platforms, and include, for example, *prompt debunking, demotion, possible removal or action against accounts* ("Tackling COVID-19 disinformation – getting the facts right", 2020)

> DEBUNKING – *Facts and logic* matter ("Identifying conspiracy theories", 2021)

Of course, conspiracists also speak of truth and facts, but their orientation towards these rationalistic values is misguided. For example, consider the scare quotes around the term 'evidence' in the first quote below and the subtle addition of the adjective 'scientific' in the following quote:

> Conspiracy theories have these 6 things in common: (1) An alleged, secret plot.; (2) A group of conspirators; (3) *'Evidence'* that seems to support the conspiracy theory; (4) They falsely suggest that nothing happens by accident and that there are no coincidences; nothing is as it appears and everything is connected; (5) They divide the world into good or bad; (6) They scapegoat people and groups. ("Identifying conspiracy theories", 2021)

> Be careful, conspiracy theories are deceptive: they *ignore scientific evidence and falsely blame individuals and groups that are not responsible for the pandemic* ("Identifying conspiracy theories", 2021)

The game is designed as a process of error elimination and thus is organised around the application of Reason to human affairs. Conspiracists are

---

[2] https://ec.europa.eu/info/live-work-travel-eu/coronavirus-response/fighting-disinformation_en

[3] https://www.who.int/news-room/feature-stories/detail/fighting-misinformation-in-the-time-of-covid-19-one-click-at-a-time

[4] https://help.twitter.com/en/rules-and-policies/medical-misinformation-policy

[5] https://www.facebook.com/formedia/blog/working-to-stop-misinformation-and-false-news

A second feature, compatible with the rationalist orientation identified above, is the inclination to work with a *binary representation* of the discourse space in which the conflict occurs. Claims are either true or false, facts are opposed to non-facts, and standpoints are either 'in' or 'out'. There is very little wiggle room. Evidential support does not come in degrees. Individuals are either on the right side of the story, or they are hopelessly mistaken. Some examples are highlighted with italics below:

> Information circulating includes *dangerous hoaxes and misleading healthcare information*, with *false claims* (such as 'it does not help to wash your hands' or 'the Coronavirus is only a danger to the elderly'). Such content is not necessarily illegal but can directly endanger lives and severely undermine efforts to contain the pandemic ("Fighting disinformation", 2021)

> To address this trend, the European Commission and UNESCO are publicising a set of ten educational infographics helping citizens identify, *debunk* and *counter* conspiracy theories. ("Identifying conspiracy theories", 2021)

> [On a section dedicated to journalists:] Emphasise *core facts, not conspiracy theories* in headlines; Reinforce core facts in the main text, using verified information ("Identifying conspiracy theories", 2021)

A third feature is that conspiracists attacking the existing consensus are not just epistemically wrong; they are *morally* wrong. This is generally because conspiracists destabilise and endanger the application of the rational solution to the problem. Conspiracy theories are morally blameable barriers concerning public goals such as health, safety, truth, trust etc. Particularly telling in this regard is the term 'infodemic', a portmanteau of information and pandemic (or epidemic), coined in the early 2000s during the SARS outbreaks. The term has been widely used for the COVID-19 conspiracism and more generally for spreading false information on various aspects of the pandemic. The World Health Organization went further and organised the "1st WHO Infodemiology Conference".[6] The semantics leave no room for ambiguity as to the interpretation of the term: false information is to truth what the viruses are to human health. For example:

> Conspiracy theories that may endanger human health, harm the cohesion of our societies and may lead to public violence and create social unrest (for example conspiracies and myths about 5G installations spreading COVID-19 and leading to attacks on masts, or about a particular ethnic or religious group being at the origin of the spread of COVID-19, such as the worrying rise in COVID-19 related anti-Semitic content) ("Tackling COVID-19 disinformation – getting the facts right", 2020)

At times, a second layer of moral culpability can be discerned. For example, conspiracy groups appear in mainstream discourse to share many of the traits we, in fact, associate with conspirators (see Sect. 6.2). This is quite ironic because conspiracists are seen as blameworthy precisely for thinking that others are forming a conspiracy. And yet, their behaviour is described as "coordinated manipulative behaviour":

---

[6] https://www.who.int/teams/risk-communication/infodemic-management/1st-who-infodemiology-conference

Platforms need to *curb coordinated manipulative behaviour* and increase transparency around malign influence operations. ("Tackling COVID-19 disinformation – getting the facts right", 2020)

We want to highlight a final feature that the monist policy response makes an effort to appear as a pluralist response that allows counter-claims and open discussion. Compared to the previous three, this last feature is more difficult to identify because we must distinguish between the spirit of the policy and its letter. The former is monist even when the latter is pluralist. Yet it stands to reason, we think, that after conspiracists are associated with 'misinformation', 'false news', 'manipulative behaviour', 'infodemic' and the like, the following are nothing but lip service appearing at the end to maintain a tolerant appearance:

All of the above remedies should be implemented in full respect of fundamental rights, in particular freedom of expression ("Tackling COVID-19 disinformation – getting the facts right", 2020)

So, what can you do? Encourage open debate and questions; Ask detailed questions about their theory in order to trigger self-reflection. [...] Don't ridicule. Try to understand why they believe what they believe. ("Identifying conspiracy theories", 2021)

After several infographics portraying conspiracism as outsiders who do not share mainstream standards of science and evidence, we cannot help thinking the advice is vacuous.

6.5 A Pluralist Alternative within Responsible Innovation?

Is it possible to approach the conflict between science and conspiracists from a pluralist perspective within the field of responsible innovation, and, if so, what would this response look like? First, we will argue that a pluralist response is possible within responsible innovation and, in fact, expected on moral and practical grounds that have spurred the field in the first place. Then we will show that a pluralist response to conspiracism can start by negating the four features of monism illustrated above.

That the field of responsible innovation is committed to some form of pluralism can be deduced both from its kinship with constructivist approaches to science such as STS and ethics of technology (Owen et al., 2013) and its aforementioned commitment to activate silent and critical publics. The pluralist suggestions advanced here will also not be strange to those who, standing on the shoulders of Weber, Berger and Latour, are quick to affirm that science is socially constructed, and thus facts are subject to discussion. And as already mentioned, the advice to turn our engagement machinery toward hard critics of science has been voiced in various ways since the field's inception (Felt & Fochler, 2010; Goodin & Dryzek, 2006; Selin et al., 2017; Wynne, 2007). Without these critical publics, including stakeholders already part of a reigning consensus would be nothing more than a self-fulfilling prophecy. The participatory process would only confirm pre-existing assumptions regarding the scientific process or product under discussion. Including critical voices is not just a good idea it is the choice that confers meaning to the inclusion process in the first place.

Has the field of responsible innovation lived up to the promise of including critics? It has to some degree, but practitioners seem primarily focused on what might be called *mild* critical sentiments. The resulting conflicts are moderate both in their force and in their effects. From selecting stakeholders to the actual involvement exercises, the process is shielded against deep disagreements between fundamentally different worldviews and interruption of already established pathways (Blok, 2019). It is sometimes said that stakeholders brought together in the dialogue on innovation must exhibit 'optimum cognitive distance' (Cuppen, 2012; Nooteboom et al., 2007). The participatory process can, in other words, be 'spoiled' if the invited stakeholders are too far away from each other (cognitively speaking). But the hypothesis of optimal cognitive distance is, in fact, developed with an eye on the company and/or group performance (Nooteboom et al., 2007). Optimal cognitive distance is needed for a group to perform well, but it can hardly be claimed that 'group performance' is the aim of participatory exercises. If the objective of the participatory game is to place a particular innovation process in context, to enrich the spectrum of perspectives on it, to broaden the central narrative around this process – in short, if responsible innovation is about *diversity* and not efficiency, then the requirement of optimal cognitive distance does not seem to apply. Indeed, the field of responsible innovation might be morally committed to seeking *maximum* cognitive distance: the parties should be just about as different as they can be (while still being part of the same world, morally speaking) in order to ensure diversity of views.

Conspiracists constitute precisely this faraway group towards which the field of responsible innovation seems committed to be responsive. But, of course, both over-inclusion (including conspiracists when it is better not to) and under-inclusion (not including conspiracists when it is better to include them) remain risks in the process. Keeping these risks in mind, we must ask whether a pluralist response to COVID-19 conspiracism can be formulated and what a pluralist response would look like. Given the four aspects highlighted in the previous section, we can formulate a provisional answer to the question: a pluralist response would amount to a negation of the four monist features identified above.

The first aspect illustrated was the rationalist narrative in which the policy response was framed. The policy exemplifies what following Blok (2019) was referred to as a 'cognitive response' to the appearance of the Other – a response that emphasises the cognitive dimension of the conflict. The policy is concentrated exclusively on facts, evidence, science etc., effectively promoting the value of truth as standing above others. A pluralist approach would not deny the importance of truth but would highlight the importance of other values, social obligations and institutions. Matters of happiness, virtues, social cohesion, justice, freedom, institutional trust, citizenship, and many more can come into play in the discussion on the COVID-19 pandemic without any one of these functioning as the central point to which all others converge.

In practice, the responsible innovation scholar would seek to create moments of interaction where all these alternative narratives can be unfolded and brought into a relationship with one another. These would be neither debunking sessions nor debates; the interaction would not be a rationalist exchange of arguments, although, of course, factuality and truth are not to be excluded from the discussion altogether. The pluralist policy or intervention can compensate for the apparent difference in authority and social status between scientists and their interlocutors. Even when the scientific community is on the right side of the argument (and there is a reigning consensus on the 'facts of the matter'), the factual aspects are but one dimension of the relationship between conspiracists and the group representing the mainstream consensus. By selecting stakeholders, topics, and interaction formats, the responsible innovation scholar can bring into practice the multi-dimensionality that a complex problem (such as a world pandemic) deserves. The conspiracists might still be wrong in the end; the scientists might still be correct. But the pluralist engagement exercise need not revolve exclusively around this one Archimedean point.

The second observed aspect is the binary representation of truth: claims are either true or false. Perhaps logical and mathematical truth must remain binary in this sense (although many-valued logics have been developed since the beginning of the 20th century). Nevertheless, the reality around complex or 'wicked' problems will resist such reduction. As Fuller puts it:

> [...] fundamental to the governance of science as an 'open society' is the right to be wrong. This is an extension of the classical republican ideal that one is truly free to speak their mind only if they can speak with impunity. [...] The underlying intuition of this social arrangement, which is the epistemological basis of Mill's *On Liberty*, is that people who are free to speak their minds as individuals are most likely to reach the truth collectively. [...] In a post-truth world, this general line of thought is not merely endorsed but also intensified (Fuller, 2018, p. 151)

Thus, even when it is true that the vaccine against COVID-19 reduces chances of hospitalization by a certain per cent, it is not clear that anything short of this statement will necessarily be false and must be discarded. After all, the statement needs to be qualified in terms of age group, method of administration, sample rates, rare but possible side-effects, statistical accuracy, epistemic assumptions etc. These variables do not altogether change the acceptability of the vaccine as a solution. Still, they do create a moral obligation to understand the broader context within which statements about the vaccine are made. In the examples given above, we have seen that policy-makers are quick to throw in terms such as 'false' and 'fake'. Still, complex situations would perhaps be better tackled with notions that allow for degrees or at least the possibility of a spectrum. There are no such things as alternative facts, but people might be more or less in the right or the wrong, people might have more or less of a point (even *malgré eux*), people might make more robust or weaker testimonies, a solution might be more or less efficient etc. There are sufficient terminologies out there to understand facts incrementally, in terms of verisimilitude, rather than categorically, in binary terms of truth versus falsity.

The third aspect referred to above was portraying those outside the consensus as morally culpable for various reasons, e.g., slowing down the roll-out of public health measures and manipulating the weak. If conspiracists appear as enemies from a forgotten tribe that has not yet enjoyed the benefits of rationalist illumination, then surely there are very few ethical demands we share with such individuals. But conspiracists are generally speaking, born and educated in the same system that is now so adamantly making a case for their moral culpability. They are part of the same social media platforms as those who defend the reigning consensus and are just as free to reject some as bogus and accept some as facts. Conspiracists might not be on the right side of truth this time, but they share with us the value of truth and other values, which means that their ethical demand for responsiveness is still valid. What does it mean to be responsive to someone who is supposedly wrong? As explained in Sect. 6.3, the responsible innovation scholar can focus on finding and formulating this common ground of core values and maintaining the agonism between the parties involved, i.e., portraying them as worthy adversaries. Argumentative discussions in various deliberative formats might be needed to find this common ground. Still, they would then function as tools for exploring the other party's views rather than testing them against the criterion of truth and Reason. The parties would debate *in order to* understand each other, not *because* they already understand each other and want to identify who's right and who's wrong.

The fourth aspect concerned the formal, tip-of-the-hat praises of pluralism and tolerance. In the current debate on COVID-19, conspiracism, pluralism, and diversity take the form of a forgotten remark at the end of an otherwise monist policy response. In this situation, the pluralist alternative would be to revert matters to their original pluralist state. In everyday life, everyone is ready to admit that there are various versions of the good life – that as a value, truth, and rationality stand alongside happiness, tradition, wealth, love, office, creativity, freedom, family, and many other more values each with its own meaning and criterion of distribution (Walzer, 1983). A pluralist response would involve taking this plurality as a starting point for the discussion rather than adding it as a supplementary remark at the end.

6.6   Conclusion

"The inclusion of stakeholders" is a phrase on many people's lips these days. The intellectual heritage of the field of responsible innovation suggests that the field is already animated to some extent by pluralism. But can this inclusive, tolerant attitude withstand the test of conspiracism? We believe it can, and we have contributed with suggestions on how this can be put into practice. In Table 6.1 below, we summarize the main differences between a monist and pluralist response to conspiracism.

Table 6.1 Monism vs pluralism in response to conspiracism

Monist response to conspiracism	Pluralist response to conspiracism
Cognitive/Rationalizing approach to conflict with an emphasis on arguments	Ethical approach to conflict with an emphasis on starting points (common ground) as well as blind spots
Binary representation of the discourse space	Non-reductionist approaches to truth
Antagonism (groups as enemies)	Agonism (groups as legitimate adversaries)
Pluralism as an end-remark	Pluralism as a starting point

The possibility of pluralism within the field of responsible innovation shows that our ideal of stakeholder inclusion does not have to leave out conspiracists, despite their radical dissensus with mainstream institutions of science (either in general or concerning a particular innovation). This does not mean, of course, that a pluralist response to conspiracism, acceptable as it might be from a responsible innovation perspective, is always the best response. There are indeed limits to inclusion set by fundamental (constitutional) rights, what Isaiah Berlin called "the great goods", and conspiracists can forfeit their right to inclusion by going against those goods. But conspiracists *qua* conspiracists appear to be 'includable' if a pluralist philosophy underpins the inclusion process. From a monist perspective, conspiracists will appear as outsiders and might be kept afar under the assumption that their views are wrong or dangerous or, in some other sense, illicit. From a pluralist perspective, accepting conspiracism as a stakeholder group and being responsive to them does not mean accepting conspiracists theories as true (or, for that matter, as false). Instead, it means following the ideal of 'science *with and for society*' to its ultimate, if unexpected, consequences.

References

Berlin, I. (1998). *The first and the last*. New York Review of Books.
Blok, V. (2019). From participation to interruption: Toward an ethic of stakeholder engagement, participation, and partnership in corporate social responsibility and responsible innovation. In R. von Schomberg & J. Hankins (Eds.), *International handbook of responsible innovation* (pp. 243–259). Edward Elgar. https://doi.org/10.1007/s13347-020-00430-7
Byford, J. (2011). *Conspiracy theories: A critical introduction. Houndmills, Basingstoke, Hampshire.* Palgrave Macmillan.
Cinelli, M., Quattrociocchi, W., Galeazzi, A., Valensise, C. M., Brugnoli, E., Schmidt, A. L., … Scala, A. (2020). The COVID-19 social media infodemic. *Scientific Reports, 10*(1), 16598. https://doi.org/10.1038/s41598-020-73510-5
Coady, D. (2006). *Conspiracy theories: The philosophical debate.* Ashgate.
Crowder, G. (2021). *The problem of value pluralism: Isaiah Berlin and beyond.* Routledge.
Cuppen, E. (2012). Diversity and constructive conflict in stakeholder dialogue: Considerations for design and methods. *Policy Sciences, 45*(1), 23–46. https://doi.org/10.1007/s11077-011-9141-7
Dentith, M. R. X. (2014). *The philosophy of conspiracy theories.* Palgrave Macmillan.

Dentith, M. R. X. (Ed.). (2018). *Taking conspiracy theories seriously: Collective studies in knowledge and society*. Rowan.

Douglas, K. M., Uscinski, J. E., Sutton, R. M., Cichocka, A., Nefes, T., Ang, C. S., & Deravi, F. (2019). Understanding conspiracy theories. *Political Psychology, 40*(S1), 3–35. https://doi.org/10.1111/pops.12568

Enders, A. M., Uscinski, J. E., Seelig, M. I., Klofstad, C. A., Wuchty, S., Funchion, J. R., ... Stoler, J. (2021). The relationship between social media use and beliefs in conspiracy theories and misinformation. *Political Behavior, 1-24*, 10.1007%2Fs11109-021-09734-6.

Felt, U., & Fochler, M. (2010). Machineries for making publics: Inscribing and de-scribing publics in public engagement. *Minerva, 48*(3), 219–238. https://doi.org/10.1007/s11024-010-9155-x

Ferrara, E., Cresci, S., & Luceri, L. (2020). Misinformation, manipulation, and abuse on social media in the era of COVID-19. *Journal of Computational Social Science, 3*(2), 271–277. https://doi.org/10.1007/s42001-020-00094-5

Fighting disinformation. (2021). Retrieved from https://ec.europa.eu/info/live-work-travel-eu/coronavirus-response/fighting-disinformation_en

Fuller, S. (2018). *Post-truth: Knowledge as a power game*. Anthem Press.

Galston, W. A. (2005). *The practice of liberal pluralism*. Cambridge University Press.

Gee, D. (Ed.). (2001). *Late lessons from early warnings: The precautionary principle 1896–2000*. EEA.

Goodin, R. E., & Dryzek, J. S. (2006). Deliberative impacts: The macro-political uptake of mini-publics. *Politics and Society, 34*(2), 219–244. https://doi.org/10.1177/0032329206288152

Groves, C. (2009). *Future ethics: Risk, care and non-reciprocal responsibility*. Taylor and Francis.

Grunwald, A. (2009). Technology assessment: Concepts and methods. In A. Meijers (Ed.), *Philosophy of technology and engineering sciences* (pp. 1103–1146). North-Holland.

Grunwald, A. (2011). Responsible innovation: Bringing together technology assessment, applied ethics, and STS research. *Enterprise and Work Innovation Studies, 31*, 10–19. http://hdl.handle.net/10362/7944

Hampshire, S. (2018). *Justice is conflict*. Princeton University Press.

Hansen, H. V., & Pinto, R. C. (1995). *Fallacies: Classical and contemporary readings*. Pennsylvania State University Press.

Hansson, S. O. (2017). *The ethics of technology: Methods and approaches*. Rowman and Littlefield International.

Harremoës, P., Gee, D., MacGarvin, M., Stirling, A., Keys, J., Wynne, B., & Vaz, S. (Eds.). (2013). *Late lessons from early warnings: Science, precaution, innovation* (Vol. 1). European Environmental Agency.

Hellstrom, T. (2003). *Systemic innovation and risk: Technology assessment and the challenge of responsible innovation*. Elsevier Ltd. https://doi.org/10.1016/S0160-791X(03)00041-1

Identifyingconspiracytheories.(2021).Retrievedfromhttps://ec.europa.eu/info/live-work-travel-eu/coronavirus-response/fighting-disinformation/identifying-conspiracy-theories_en

Jasanoff, S. (2003). *Technologies of humility: Citizen participation in governing science*. Kluwer Academic Publisher.

Kekes, J. (1993). *The morality of pluralism*. Princeton University Press.

Larson, H. J. (2018). The biggest pandemic risk? Viral misinformation. *Nature, 562*(7726), 309–310. https://doi.org/10.1038/d41586-018-07034-4

Lassman, P. (2011). *Pluralism*. Polity.

Lee, J. D. (2014). *An epidemic of rumors: How stories shape our perception of disease*. University Press of Colorado.

Mouffe, C. (2005). *The return of the political*. Verso.

Naeem, S. B., Bhatti, R., & Khan, A. (2021). An exploration of how fake news is taking over social media and putting public health at risk. *Health Information and Libraries Journal, 38*(2), 143–149. https://doi.org/10.1111/hir.12320

Nagel, T. (1986). *The view from nowhere*. Oxford University Press.

Needleman, H., & Gee, D. (2013). Lead in petrol 'makes the mind give way'. In EEA (Ed.), *Late lessons from early warnings: Science, precaution, innovation* (Vol. 1, pp. 46–79). European Environmental Agency.

Nooteboom, B., Van Haverbeke, W., Duysters, G., Gilsing, V., & Van den Oord, A. (2007). Optimal cognitive distance and absorptive capacity. *Research Policy, 36*(7), 1016–1034. https://doi.org/10.1016/j.respol.2007.04.003

Oakeshott, M. (1991). *Rationalism in politics and other essays*. Liberty Fund.

Owen, R., Stilgoe, J., Macnaghten, P., Gorman, M., Fisher, E., & Guston, D. H. (2013). A framework for responsible innovation. In R. Owen, J. Bessant, & M. Heintz (Eds.), *Responsible innovation* (pp. 27–51). Wiley.

Paxton, M. (2020). *Agonistic democracy: Rethinking political institutions in pluralist times.* Routledge.

Popa, E. O., Blok, V., & Wesselink, R. (2020a). An agonistic approach to technological conflict. *Philosophy and Technology, 34*(4), 717–737. https://doi.org/10.1007/s13347-020-00430-7

Popa, E. O., Blok, V., & Wesselink, R. (2020b). A processual approach to friction in quadruple helix collaborations. *Science and Public Policy, 47*(6), 876–889. https://doi.org/10.1093/scipol/scaa054

Pummerer, L., Böhm, R., Lilleholt, L., Winter, K., Zettler, I., & Sassenberg, K. (2020). Conspiracy theories and their societal effects during the COVID-19 pandemic. *Social Psychological and Personality Science, 19485506211000217*, 10.1177%2F19485506211000217.

Romer, D., & Jamieson, K. H. (2020). Conspiracy theories as barriers to controlling the spread of COVID-19 in the U.S. *Social Science and Medicine, 263*, 113356. https://doi.org/10.1016/j.socscimed.2020.113356

Selin, C., Rawlings, K. C., de Ridder-Vignone, K., Sadowski, J., Altamirano Allende, C., Gano, G., … Guston, D. H. (2017). Experiments in engagement: Designing public engagement with science and technology for capacity building. *Public Understanding of Science, 26*(6), 634–649. https://doi.org/10.1177/0963662515620970

Stilgoe, J., Lock, S. J., & Wilsdon, J. (2014). Why should we promote public engagement with science? *Public Understanding of Science, 23*(1), 4–15. https://doi.org/10.1177/0963662513518154

Stirling, A. (2008). "Opening up" and "closing down" power, participation, and pluralism in the social appraisal of technology. *Science, Technology, and Human Values, 33*(2), 262–294. https://doi.org/10.1177/0162243907311265

Tackling COVID-19 disinformation – getting the facts right. (2020). *[Press release]*. Retrieved from https://eur-lex.europa.eu/legal-content/EN/TXT/PDF/?uri=CELEX:52020JC0008andfrom=EN

Timmermans, J., & Blok, V. (2018). A critical hermeneutic reflection on the paradigm-level assumptions underlying responsible innovation. *Synthese, 198*(19), 4635–4666. https://doi.org/10.1007/s11229-018-1839-z

van den Hoven, J., Doorn, N., Swierstra, T., Koops, B., & Romijn, H. (Eds.). (2014). *Responsible innovation 1: Innovative solutions for global issues.* Springer Science + Business Media Dordrecht.

Walzer, M. (1983). *Spheres of justice: A defense of pluralism and equality.* Basic Books.

Wilsdon, J., & Willis, R. (2004). *See-through science: Why public engagement needs to move upstream.* Demos.

Wynne, B. (2007). Public participation in science and technology: Performing and obscuring a political–conceptual category mistake. *East Asian Science, Technology and Society: An International Journal, 1*(1), 99–110. https://doi.org/10.1215/s12280-007-9004-7

Open Access This chapter is licensed under the terms of the Creative Commons Attribution 4.0 International License (http://creativecommons.org/licenses/by/4.0/), which permits use, sharing, adaptation, distribution and reproduction in any medium or format, as long as you give appropriate credit to the original author(s) and the source, provide a link to the Creative Commons license and indicate if changes were made.

The images or other third party material in this chapter are included in the chapter's Creative Commons license, unless indicated otherwise in a credit line to the material. If material is not included in the chapter's Creative Commons license and your intended use is not permitted by statutory regulation or exceeds the permitted use, you will need to obtain permission directly from the copyright holder.

Chapter 7
Values as Hypotheses and Messy Institutions: What Ethicists Can Learn from the COVID-19 Crisis

Udo Pesch

7.1 Introduction

The global crisis that emerged from the COVID-19 pandemic created profound moral challenges. The crisis forced us to think about what is really important concerning individual values and societal ones. For example, trade-offs needed to be made between the value of human life and the value of the economy. In addition, questions need to be answered about how society could deal with vulnerable people, how sick ones could die humanely, and how the deceased could be given a dignified farewell (e.g. see Centrum voor Ethiek en Gezondheid, 2020).

The academic field of ethics has contributed substantively to tackling these challenges regarding the evaluation of practices and policies and the determination of the correct values (Dartnell & Kish, 2021; Kim & Grady, 2020; Verweij et al., 2020). Such evaluations usually aim to clarify ongoing discussions about policy measures, by isolating the underlying problem from its political and societal contexts. This helps to understand the underlying moral structure of the problem at hand, but in many cases moral problems are created by the institutional arrangements set up to pursue public values and the workings of the public debate which determines these values. These closely related issues cannot be straightforwardly isolated from the practices and policies subjected to ethical evaluation.

To unpack this claim, we can look at the critical role of the division of modern society into a public and a private sphere. This division allows us to maintain personal values that guide individual choices, while a collective course can be determined based on the idea that there is a common ground which is based on shared values. The values that are to be shared are decided upon within deliberative processes that are based on 'public reason', in which individuals develop positions

U. Pesch (✉)
Delft University of Technology, Delft, The Netherlands
e-mail: U.Pesch@tudelft.nl

© The Author(s) 2022
M. J. Dennis et al. (eds.), *Values for a Post-Pandemic Future*, Philosophy of Engineering and Technology 40, https://doi.org/10.1007/978-3-031-08424-9_7

based on their idea what is best for the social collective they are part of. With that, the separation between a public and a private sphere can be seen as a basic institutional structure, designating specific rules and norms to different social contexts. This institutional structure encompasses a number of other institutions that allow a collective course to become effectuated, such as parliamentary democracy and government.

The COVID-19 crisis testifies that these institutions and the substantiation of public reason are not without problems. To start with, the state apparatus that has been developed to pursue public goals is often subject to bureaucratisation. For instance, the organisational processes that have been set up to distribute vaccines were so rigid that medicines had to be thrown away, leading to frustration among doctors, politicians, and the public (March, 2021). Moreover, in some cases, the rules that have been implemented to safeguard public health are now blamed for endangering public health.

The pursuit of public values is organised in national contexts, while COVID-19, following the globalised socio-economic system, is not restricted by any border (Ludovic et al., 2020). This, for instance, has meant that expats and tourists ended up in isolation, far away from friends and family, while having no opportunity to meet people in real life. It also meant that vaccines are spread unevenly worldwide, which opens the chance for new virus variants to develop in regions where vaccination rates are still low.

We can also have a look at the widespread distrust in the pharmaceutical industry. It is typical of this branch that it crosses the boundaries between the institutional realms of the state, the market, and science. These realms respectively pursue the public of healthcare, business enterprise, and creating new knowledge. Big pharma combines these goals, which creates moral concerns as companies make enormous profits due to novel medicine that the majority of the population needs to take to reduce the impact of the COVID-19 crisis. Moreover, the amount of financial gains makes the industry susceptible to critique and conspiracy thinking. After all, it will be in its interest to sell vaccines, even if they do not work effectively or may have adverse side effects.

It is also hard to recognize public reason in discussions that are held on social media like Twitter and Facebook in the shaping of a public discourse. These social media are often taken as a platform (or even *the* platform) that allows for a public debate. However, it is incorrect to assume that these media facilitate such a debate. Firstly, many activities are initiated by non-humans, such as bots that aim to distort the debate or organisations that seek to further organisational interests. As such, media do not allow individual citizens to the exchange their perspectives on what should be the 'public' interest. Instead, internet discussions tend to be partisan, skewed and distorted, contributing to further discontent and distrust (see Steinert, 2020).

Such developments contribute to the rise of a vocal group of people that do not want to be vaccinated or resist COVID-19 measures. Some voices even do not want to align with public reason in the first place. There are political parties and protest groups that explicitly contest the legitimacy of political institutions to serve the public. As such, the very idea of a common ground is challenged, threatening our capacity to find shared values by public reason.

These examples show that the constitutive relation between institutional and public reason is subject to two problems. First, the boundary between publicness and privateness can be understood in a variety of ways, which may contribute to societal worries. Second, there is no straightforward way to establish shared values, as in reality this is rather a disorganised process in which the idea of a common ground cannot be taken for granted.

To deal with these issues, I will develop an approach in this chapter that allows the *translation* of ethical considerations into institutional arrangements and the *structuring* of societal processes that give rise to public opinion. At its core, this approach aims to bring politics and ethics more closely together than is currently the case in ethics and social research.

In this, I take the COVID-19 crisis as an episode that allows us to learn about the moral role that institutions play in safeguarding relevant values, while these institutions simultaneously both shape and are shaped by public deliberations. In other words, this chapter will not ask how ethics can help us cope with the COVID-19 crisis (which, without any reservations, is still a cardinal question), but it will reverse this question and ask what *ethics can learn* from the COVID-19 crisis.

This chapter proceeds in the following way. In Sect. 7.2, I will discuss the opposition between politics and ethics in ideal-typical terms. This opposition holds that there are two orientations: either values are fully *independent* from politics, as can be recognised, at least to some extent, in analytic ethics, or values are fully *dependent* on politics, which appears to be a common assumption within constructivist philosophy and social research. Both orientations are counterproductive as politics *and* morality play a crucial role in social life.

In Sect. 7.3, I will first turn to ideas developed by John Dewey, Jürgen Habermas, and Charles Taylor to develop a more productive approach. Their ideas allow for an account in which values are not static and in which values are not isolated from politics or public discourse. However, in their work on deliberative democracy, these authors still seem to regard institutions as instrumental to the moral organisation of society. The conditions of deliberative democracy are mainly described as theoretical constructs, which goes to the extent of the attention for the volatility of the relation between values and institutions.

In Sect. 7.4, I will explore an approach that is able to understand the workings of institutions in such a way that they still allow for the pursuit of a society in which relevant values are safeguarded and pursued. This approach sees values as 'moral hypotheses' that are tested and substantiated in the institutions that characterise modernity. In other words, modern institutions allow us to find out what values mean in real-life contexts, potentially giving rise to demands to reconceptualise these values or adjust these institutions. At the same time, the moral problems caused by institutions themselves are not sufficiently subjected to theoretical reflection.

In the final section of this chapter, I will reiterate some of my central claims. I will restate their importance given other global challenges we are currently facing, with climate change being the most critical of these.

7.2 An Opposition Between Politics and Ethics

In this section, I will depict an opposition between two orientations that are recognisable in the literature. On the one hand, there is 'analytic ethics', which appears to be the dominant approach to philosophy and ethics in academia (Bell et al., 2016). It is mainly within this approach that one can recognise the tendency to exclude politics as a relevant factor. Contrastingly, there is the approach developed by those that can be called 'deconstructivist' scholars where the independent status of moral issues is discarded.

It needs to be admitted that the description of this opposition has an ideal-typical nature; in real life, it is hard to pinpoint the analytical and deconstructivist orientations and most scholars entertain positions that are much more nuanced. Still, these orientations guide such scholars in their epistemological and methodological assumptions. Thus, they figure as ideals that are not necessarily followed but may also figure as the background against which alternative approaches are developed. One way of explaining the ideal-typical opposition between these two approaches is to refer to Kant's distinction between the analytical and synthetic a priori. While ethicists tend to position moral issues in the domain of the analytical a priori or take this as the starting point from which they deviate, the deconstructivist approach takes moral issues to belong to the synthetic a priori.[1]

Many analytic ethicists are concerned with the question of the universal and objective validity of moral claims. In this, the grounds that allow the justification of moral claims is widely contested (Roeser, 2005), giving rise to a variety of meta-ethical positions. What characterises most positions however is that 'moral truths' rely on their conceptual consistency and the eradication of contingent factors (cf. Erdur, 2016). Politics and power are among the most salient contingent factors. As such, they are non-essential phenomena that should not be considered to find out what is morally relevant (cf. Brink, 1989). Values are then easily perceived as unchangeable moral truths; in this, ethicists have taken up the task of discovering these (cf. Korsgaard, 2015). This means that discussions about the relevant values for specific societal challenges lose their relevance for ethics. Another element of analytic ethics that plays a decisive role here is the focus on individual agents vis-à-vis society as a whole. Individuals are taken to be autonomous, having no intrinsic

---

[1] My analysis is to a large extent based on academic experience in different scholarly fields. In the last two decades I have worked in policy studies, STS, and ethics of technology and I was often puzzled by the underlying normative assumptions of this fields, as these are usually not made explicit. The account of this section can be seen as an attempt to identify these assumptions, in which literature has been used to systemise these observations.

moral connection with their communities (Pesch, 2020a). Again, there are many that authors do not maintain such far-reaching presumptions about the status of moral claims. However, it is rather typical that these authors explicitly contrast themselves with these 'realist' starting points in order to present their own arguments – showing the relevance of the orientation I sketched out here.

For the deconstructivist approach, politics and power figure very much as its key concern. In this approach, the assumption that moral claims can be tested as if they were truth claims is disputed: moral claims are not considered to have objective status. Instead, they are derived from concrete social structures. This means that it cannot be expected that a moral claim will have the same status independent from time and place. The origins of this approach may be found in Nietzsche's work and are particularly recognisable in poststructuralist sociology, critical theory (Hoy, 2005) and STS (Sismondo, 2008). Starting point in all of these branches of research is that every truth is a human-made truth, with power structures, belief systems, and ingrained routines determining what is true and right. In other words, the question about what is true is a political question (Foucault, 1997). When claims are studied that are considered moral 'truths', one should not look at the content of these claims but at the social and political context within which these claims are made. Moral truths must be deconstructed to reveal which social conditions ensure that a particular morality is maintained as true.

The deconstructivist approach provides a sharp critical toolkit that might be used to identify those moral claims that are taken as moral truths. As such, moral wrongs could be discovered and, potentially, strategies to overcome these wrongs could be formulated. At the same time, however, the approach is *methodologically nihilistic*. The approach can be used to uncover moral wrongs, but it does not provide the tools to say why something is wrong and what can be done to make things right. Often this methodological nihilism spills over into moral relativism, which means that *no* moral system is deemed better than another because moral standards used to say that something is 'better' are also culturally embedded. Strategies to overcome moral wrongs then become senseless. Interestingly, research domains closest to the deconstructivist tradition such as critical theory, STS, and feminist studies have strong normative outlooks, endorsing justice and engagement of scholars (for instance see Mamo & Fishman, 2013), while genuinely nihilistic positions tend to spill over to fields such as policy sciences. In this field, moral issues are often exclusively taken as mere stakeholders' input, meaning that this is empirical not normative information. It is telling that in policy sciences, politics is often portrayed as a 'game'(cf. Scharpf, 1997), a process that is considered amoral in itself. The game is played by actors who aim to optimize their goals that are perceived as subjective choices that need no further scrutiny.

But the fact that different value systems coexist (and that it cannot be stated which of those systems is better) does not imply that value systems in themselves have no value (Roeser, 2005). In the end, we cannot help being moral beings: *every* statement we make about how we relate to others is an inherently moral statement (also see Pesch, 2020). All our choices and assessments invoke normative qualifications about what is 'better' or what is 'worse'. Moreover, we are communal beings, in which the political realm is where we can decide what we find important as a society, and it is also the realm that allows society to organise itself as a moral community (cf. Arendt, 1958). Politics *is* the organisation of ethics at the collective level.

Indeed, the COVID-19 crisis brings about a range of ethical choices that need to be settled collectively. It made clear that we do not live in a nihilistic universe; instead, we are apprehensive about older adults who die in an inhumane way – even if we do not know these people personally. Most of us have seem to have sacrificed our daily routines and our direct interests for the common good without hesitation (Lynch & Khoo, 2020). It shows that humans have the innate quality to help out people in need (Warneken & Tomasello, 2006). Also those who fall outside every risk group know how to empathise with the victims, their families and the care providers, testifying the claim that empathy is a key driver for human behaviour (De Waal, 2006; Tomasello et al., 2005). Having said that, the pandemic is in its very nature a 'collective action' problem (Harring et al., 2021): the ethical questions that are brought about by COVID-19 cannot be solved by the aggregate of individual reactions. This raises the question of how society can organise itself to make collective choices in the face of a pandemic. What are the conditions that permit politics to focus on the values that we collectively consider to be important?

To answer this question, it is necessary to see what relevant values are and to understand how these can be substantiated in practical arrangements. Here I start with the Dewey's pragmatist approach to values, which allows a non-static account of values that can be straightforwardly embedded within the idea of deliberative democracy. However, as I will explain below, theories on deliberative democracy tend to underplay the complexity of institutional arrangements: while values are not seen as static any longer, institutions still are. To overcome this problem, it is vital to be able to critically deconstruct the workings of institutions to see whether institutions can deliver their intended functions. After all, they should be adjusted if they do not do so.

7.3   Values and Deliberative Democracy

To allow the connection of values and politics, I will first turn to *deliberative democracy*, which can be seen as the political shape that will enable us to say something about how we want to live together, by determining the values that we find worth pursuing as a collective. In this, we can understand values as the following: values inform the *understandings* that allow people to make sense of social phenomena to make decisions in anticipation of future events (Dewey, 1922, 1927). This account of values assumes that humans constantly judge what to do next by interpreting situations in terms of whether they are desirable or undesirable. Values give normative significance to a broader range of experiences and projections. They can be considered concepts that *aggregate* a variety of impressions that allow agents to prepare for future actions (Habermas, 1985; Rawls, 1997). As a higher-order categorisation of meanings, values *can be made explicit*, enabling them to be the basis for collective deliberation and decision-making processes. It is the fact that values can be made explicit that allows the *deliberate organisation of society*, facilitating discussions about collective courses of action. In other words, the explication of values and the deliberation on their prioritisation provides effective and legitimate forms of collaborative decision making.

This is done by enacting public reason, individuals in their capacity as citizens engage in deliberation about what is good for society as a whole (Habermas, 1985; Rawls, 1997). Thus, individuals imagine themselves as part of a greater whole, members of a public, which motivates them to contribute to discussions about the greater good. Subsequently, the outcomes of these discussions are taken as guidance to establish and adjust institutions, which ensures that these institutions are responsive to society (cf. Taylor 2002). In other words, by making values explicit and turning them into objects of collective reflection, we can make justify them according to moral standards that, likewise, have been subjected to collective scrutiny and deliberation.

The deliberative organisation of society is mainly done by erecting the relevant institutions and institutional domains. These institutions and domains compartmentalise social reality into different social contexts in which specific rules are maintained. They allow society to structure itself according to the right moral standards so that essential values can be maintained and pursued within social collectives.

Maybe the most basic compartmentalisation is the separation of social reality into a public and a private sphere. This construction of a boundary between these two spheres allows for both 'negative' and 'positive' freedom (see Berlin, 1969). While negative freedom is secured by installing a private sphere in which an individual is not restricted by others in her activities; positive freedom', understood as the freedom for collective self-determination, is pursued by the establishment of a public sphere in which members of a society can settle on a collective course by political action. Political theory presents the strict division into a public and private sphere as the way to overcome the problem of *value pluralism*, which is an elementary problem for democracy because even if there is consensus about the prominence of certain values, individual persons will diverge in their assessment of the relative importance between them and their understandings of these values. As authors like Jürgen Habermas (1996), John Rawls (2009), and Richard Rorty (1989) maintained, the public sphere allows for a common ground at the collective level, while individuals are entitled to their own sets of moral preferences in the private sphere.

The COVID-19 pandemic shows that the boundary between the public and private sphere is far from unequivocal. While a majority of people in developed countries comply with governmental measures, such as lock-downs and vaccination programmes, a minority opposes these measures, sometimes in very vocal ways. To a significant extent, this opposition is the result of people who contest the boundary that is drawn between the private sphere and the public sphere plays, but it is also the result of the indiscriminate use of different conceptions of this boundary.

Negative freedom informs a first conception of the public/private distinction in which the private sphere allows for choices about how to live, where to go, with whom to interact. We use this freedom to shape who we are, to constitute our personal identity. Public authorities may enforce laws and policies that ensure justice and well-being, but challenging the liberty to shape one's identity is not acceptable in any democracy. Opponents to corona-related measures as lockdowns perceive these measures to form (or threaten to form) such a challenge.

However, one may also recognize other conceptions of the public/private boundary that play a role in the societal controversies concerning COVID-19. An epidemic challenges the boundaries between individual and collective in a physical manner. This is especially relevant, as the integrity of the individual body can be said to be the epitome of the private sphere: the core of privacy is that we aim to hide uncontrollable bodily functions from the sight others (Moore, 1985; Pesch, 2015). In fact, this physical understanding of the public/private boundary that revolves around autonomy, control and dignity underlies the political understanding that pertains to freedom. A virus invades the individual body without respecting its integrity at all, but it does not challenge our understanding of the public/private boundary because of its invisibility. The injection of a vaccine however clearly breaches the boundary between what is inside and what is outside of the body. Only other people in whom we have special relation, our family or trusted professionals such as doctors, are allowed to cross this boundary. Even though in the case of COVID-19, the integrity of the body is compromised by medical specialists, this action is still commissioned by the government, which ought to refrain itself from intervening in the private sphere of the body.

This bodily connotation of privateness and publicness transfers into an informational connotation. Just like we need to keep control over our bodily processes, in the sense that we hide these processes from the sights of others (Geuss, 2001), we also need control over what others know about us. Debates about data privacy are derived from this need to keep things secret. Apps that track the movements of people or that show whether a person is vaccinated or not, may violate this need.[2] Especially as the combination of different data streams may be used to construct a profile a person, severely reducing the ability of that person to control what she wants to keep secret.

Publicness and privateness not only pertain to the relation between an individual and a social collective, this conceptual pair is also used to separate institutional domains that allow us to categorise roles of individuals and organisations. For instance, the market domain is portrayed as a private domain as well, which grants businesses and consumers the freedom to pursue their preferences by exchange goods and services for money. Measures to reduce the number of human contacts in public places precisely target private enterprises as stores, restaurants, or music venues are closed, challenging the freedom that characterises the market as a private domain. Moreover, representatives of these enterprises feel that they are now responsible for solving a problem for which they are not responsible: they have not

---

[2] https://www.aldeparty.eu/corona_dictatorship_watchdog, accessed on 10-12-2021.

caused the pandemic and they are not the designated party to do something about it; this is considered a task for the government or individuals.

The idea of bodily integrity also plays a role on another level. A nation is often portrayed as a 'body' that needs to be protected against invasive forces. The metaphor of viruses and disease is often used, and mostly abused, to demonise and exclude strangers (Bauman, 2013). In the case of COVID-19, we are dealing with a real virus of which the spread is not restrained by the borders of a nation, just like the virus is not restrained by the boundaries of the human body. Reactions to fight the virus predominantly have a domestic character and there seems to be a lack of international coordination. There is no global public sphere that allows for collective self-determination, instead there are organisation such as the World Health Organization (WHO), which are very much confined by national interests and outlooks (Davies & Wenham, 2020).

Individuals appear to have many reasons not to get vaccinated controversies (Murphy et al., 2021), which may pertain to the different conceptions of the public/private boundary levels. In the end, this boundary is so elusive that it is untenable to maintain a neat separation into two spheres (cf. Benn & Gaus, 1983; Weintraub, 1997). Another related problem that can be recognised in discussions about corona-related measures is that there is no 'agreement to disagree', a principle that can be seen as the foundation of deliberative democracy. Not only the legitimacy of public authorities is disputed by some, with populist leaders and protestors speaking of dictatorship and even genocide,[3] also the very idea that we share the same reality appears to be disputed as Bruno Latour argued (2013). Latour wrote about climate change, but his words also match the refutations of scientific findings about the existence or seriousness of the coronavirus (Hartwell & Devinney, 2021).

What these examples derived from the COVID-19 crisis show is that values can mean different things between different social contexts, but even they can mean different things *within* a singular context. Moreover, these meanings can always be subjected to societal contestation. This is illustrated by many faces of the public/private boundary that play a role in debates about corona-measures. This boundary outlines a compartmentalisation into different institutional spheres that allow certain values to be pursued, however, this boundary can be drawn in a variety of ways, invoking different understandings of values. The articulation of values via public reason can be taken as a theoretical construct that figures as a normative ideal that guides the further development of the institutions that support deliberative democracy, but to take further steps in de development of deliberative democracy, the black-box institutions need to be opened.

---

[3] E.g. https://www.rfi.fr/en/france/20210715-down-with-dictatorship-anger-at-france-s-sweeping-new-covid-rules-macron-vaccination-yellow-vests; https://theconversation.com/bolsonaro-faces-crimes-against-humanity-charge-over-covid-19-mishandling-5-essential-reads-170332   accessed   on 18-11-2021.

7.4 An Enlightened Moral Project and Messy Institutions

This section will propose a tentative approach that integrates the attention for values, institutions, and public deliberation. This approach begins by acknowledging the cardinal role of public values developed and entertained in deliberative processes, and then proceeds by acknowledges the conditionality of institutional developments.

Indeed, processes of deliberation should not be seen as isolated from the *institutions* that support them. The institutions developed in modernity embody and reproduce certain values that cannot be discarded if one wants to maintain democracy. Political institutions such as parliamentary democracy, the legal system, and public administration, and the non-political institutions of the capitalist market and modern science to facilitate values like autonomy, freedom, justice, dignity, well-being, and progress in a myriad of ways. For instance, we can see how freedom is shaped differently in a political or in an economic context: while parliamentary democracy allows for political freedom by giving citizens the right to vote, to join a political party, to run for office, etc., the capitalist market allows for economic freedom by giving consumers the right to buy the products or services they prefer.

As such, institutions figure as the vehicle with which society has been organised to enable collective processes of moral deliberation. To underpin my analysis, I'd like to portray the modern era that emerged with the Enlightenment here as a 'moral project': the collective attempt to actively shape the world according to given moral hypotheses. This moral project revolves around the establishment and further adjustment of institutions, which can be defined as societal contexts in which given sets of rules guide collective action. Values can be taken as 'moral hypotheses', they evolve from public deliberation and then are further articulated and tested within institutions. Institutions allows us to find out what a value actually means within a certain setting.

In this, an ongoing dialectical relation can be recognised: on the one hand, the right institutions are set up following collective deliberation; on the other hand, the capacity for public reason is nurtured by setting up the right institutions. Over the course of the Enlightened moral project, a patchwork of institutions has been development that have led to the compartmentalisation of society into different contexts that secure and further shape certain values. The further evolution of this patchwork of institutions is characterised by dynamics that play out within and between institutions. These dynamics are messy and they have a major impact on the further substantiation of values, compromising the capacity of institutions to test values as hypotheses.

A first issue is that institutions tend to create specific path-dependencies, most notably the rules that characterise an institution often come to form a reality on their own. This process of bureaucratisation means that goals are 'displaced', and rules that have been set up to safeguard certain values may become ends in themselves (Merton, 1940). Also in the context of corona-measures processes of goals displacement might become a reality, for instance when tracing apps will continue to be used after the end of the pandemic to serve other goals, such as surveillance. But such processes of goal displacement may also be traced at the level of institutional domains (Pesch, 2014), which give rise to societal distrust. For instance, looking at the political domain, we see how political parties have become subject to Michels's iron rule of oligarchy (Michels, 2019); politicians are recruited from a narrow societal segment and stick to party discipline. Political agendas are to a large extent set by media hypes and lobbyists pursuing a specific interest (Lowi, 1969)). Also in the market domain, a decline of responsiveness to its original goal can be observed. Companies – especially the larger ones – are often more reactive to the wishes of shareholders than to the wishes directly expressed by consumers or groups of consumers, so that the freedom of individual consumers to pursue their self-interest is seriously obstructed (Galbraith, 1998; Mazzucato, 2018). Likewise, the domain of science reveals patterns of institutional goal displacement, for instance in case of the emphasis on quality measures such as impact factors and past track record, which brings about certain problems, such as the hampering of scientific activities that do not belong to the dominant paradigms (Macdonald & Kam, 2007).

The second issue is that the *boundaries* between institutions are usually ill-defined. At these boundaries, there are continuous negotiations about which rules are valid on which occasions. For example, one can think of the boundary between political decision-making and science-based expertise (also see Lindblom & Cohen, 1979). The demands for accurate science and effective democratic decisions may be conflicting. What counts as 'good' science may not be 'good' decisions and vice versa. Workable solutions are established, but these solutions are temporary and conditional (Gieryn, 1983; cf. Jasanoff, 1990). In Dutch policy-making, science-based knowledge is coordinated by the National Institute for Public Health and the Environment (RIVM) (Pesch et al., 2012). This 'boundary organisation' gathers scientists and medical expertise and gives advices to the government.[4] A first problem is that an ongoing pandemic brings about many questions that cannot be answered yet by science. As the advice given is based on knowledge that is incomplete it can easily be contested by other scientists. In the Dutch debate, alternative interpretations were quickly distributed via traditional and social media. The proliferation of such interpretations appeared to have undermined the credibility of the advice and eventually also the measures that are taken. A second problem is RIVM has been accused by other scientists, politicians, and by the general public that is was doing politics instead of science, by being too close to the policy domain.[5] A third problem is that science-based knowledge, even if it is complete, cannot serve as the exclusive ground for political decisions, as these have a moral and not a factual character. In the end, the government has to determine which measures have to be taken, by making trade-offs between competing values in a situation that is highly uncertain. As Prime Minister Mark Rutte expressed: "In crises like these, you

---

[4] https://www.rivm.nl/coronavirus-covid-19/omt, accessed on 20-11-2021

[5] For instance see, https://www.medischcontact.nl/nieuws/laatste-nieuws/artikel/kritische-hoogler-aren-vinden-de-wetenschappelijke-basis-van-coronamaatregelen-onhelder.htm, accessed on 20-11-2021

have to take 100% of the decisions with 50% of the knowledge and bear the consequences of these decisions".[6]

The variety of institutions brings about a compartmentalised social life granting
certain roles to actors who represent certain institutions, but in a situation that is as
complex as COVID-19, this compartmentalisation cannot be maintained. Scientific
experts are asked to give advice on the political decisions that need to be taken. We
allow doctors to intervene with our bodily integrity, but vaccination programmes
have to be incited by government. Pharmaceutical industry needs to have to have the
incentives to innovate that of a competitive market provides, so the appropriate
medicine is developed. The mixture of activities from different institutional domains
raises discontent, but it a discontent that we have to accept in order to address major
societal challenges. In the end, there is no singular solution to these problems, science and politics are institutions that serve their own values and have their own rules
and bringing them together will inevitably give rise to contestation.

The assignment for theorists of deliberative democracy is not to find out the conditions that would give rise to full consensus about the status of values or to create
arrangements that are 'perfect' in theory. The take on values as hypotheses shows
that institutions are contexts that allow for *moral learning*, for ways of finding out
what values mean or can mean in specific contexts. In this, societal conflict can be
seen as a necessary source of information (Cuppen, 2018; Rip, 1986), it shows
whether values substantiated in institutions align with societal specifications of
these values. It is further input of a messy deliberative process that may give rise to
the adjustment of institutions (Callon, 1998; Pesch, 2021). By all means, COVID-19
has given rise to a global crisis, but it is not necessarily a democratic crisis (Walby,
2021). Because of the limitations of time, societal debates are heated, chaotic and
sometimes nasty, but it we accept them as learning opportunities, then we can use
them to move forwards. This does not mean that we should accept nor underesti-
mate active attempts to cause a rift in the idea of a common moral ground or a
shared reality as undertaken by some populist leaders. Such attempts are no mere
expressions of the freedom of speech or the freedom to have political preferences;
on the contrary, they threaten these freedoms as they aim to undermine the democratic institutions that serve them.

7.5 What Can Ethics Learn from the COVID-19 Crisis?

Let me use this final section to recap the central claims that have been made in this
chapter. This chapter proposed values as concepts with which we give a multitude
of situations moral significance, with which we can determine whether we find
something good or bad, and with which we can compare certain conditions in

---

[6] https://nos.nl/video/2326873-rutte-we-hebben-iedereen-nodig-17-miljoen-mensen, accessed on
20-11-2021

normative terms. Specifying a value is not an end in itself but guides actions and choices. A value is an evaluation that motivates an action or an intervention. This means that an ethical analysis must not only determine whether something is 'right' or 'wrong', but especially *how* it can be improved. It concerns the identification of alternative options for action and the exploration of the consequences of those options. Indeed, values can be seen as conceptual means to organise our lives morally – either at the individual level or at the level of the social collective.

In this, deliberative democracy appears to be the most suitable form for the moral organisation of social collectives. At the same time, it needs to be acknowledged that decision-making processes within deliberative democracy are messy. The institutions set up to serve and protect us are not without problems, and they should constantly be subjected to reconsideration and redesign. Ethics can nurture the dialectical relation between institutions and values by forwarding and fine-tuning the moral hypotheses about how specific understandings of values can be secured in real-life institutions. Here, deconstructivist methods – dropping their nihilism – can be applied, not to denounce the reality of values but to unravel moral claims with which further moral refinement can be pursued. Also, we have to have more thorough accounts of how societal contestation can be used to adjust further institutional development, so to warrant their function as contexts in which values-as-hypotheses can be tested. The grounds and conditions for moral learning ought to be mapped out in much more detail.

The need to have an approach that integrates reflections about values, institutions, and deliberation is pressing, as COVID-19 is far from the only global crisis that invokes firm value-laden decisions we are facing. One only has to think about the radical societal, political, and moral changes needed to take on the challenge of global climate change (Pesch, 2018, 2020b). For example, it is easy to consider the COVID-19 pandemic to be a 'wasted' crisis. The arrival of COVID-19 could have been taken up as an opportunity to reconsider the lock-in of vested interests and incumbent practices; instead, policy decisions primarily served the continuation of the existing economic status quo, urging producers to produce more and consumers to consume more (cf. Dartnell & Kish, 2021; Heintz et al., 2021). Not only does this reveal the reproduction of incumbent institutional practices, but it also testifies the unwillingness of political leaders to address issues that give rise to societal contestation. This suggests that the value pluralism that should be key to public deliberation is seen as unwanted (also see Cuppen & Pesch, 2021; Pesch & Vermaas, 2020). Hopefully, an ethics that can pinpoint these shortcomings and give concrete recommendations would help conquer such developments. Though I have only explored my ideas tentatively in this chapter, I think that seeing messy institutions as contexts within which values can be tested and substantiated as hypotheses would serve such an ethics.

References

Arendt, H. (1958). *The human condition*. University of Chicago Press.

Bauman, Z. (2013). *Modernity and the holocaust*. Wiley.

Bell, J. A., Cutrofello, A., & Livingston, P. M. (2016). *Beyond the analytic-continental divide*. Routledge.

Benn, S. I., & Gaus, G. F. (1983). The public and the private: Concepts and actions. In S. I. Benn & G. F. Gaus (Eds.), *Public and private in social life* (pp. 3–27). Croom Helm.

Berlin, I. (1969). Two concepts of liberty. *Berlin, I*, 118–172.

Brink, D. O. (1989). *Moral realism and the foundations of ethics*. Cambridge University Press.

Callon, M. (1998). An essay on framing and overflowing: Economic externalities revisited by sociology. *The Sociological Review, 46*(S1), 244-269. 10.1111%2Fj.1467-954X.1998.tb03477.x.

Centrum voor Ethiek en Gezondheid. (2020). *Ethiek in tijden van Corona*. Centrum voor Ethiek en Gezondheid.

Cuppen, E. (2018). The value of social conflicts. Critiquing invited participation in energy projects. *Energy Research and Social Science, 38*, 28–32. https://doi.org/10.1016/j.erss.2018.01.016

Cuppen, E., & Pesch, U. (2021). How to assess what society wants? The need for a renewed social conflict research agenda. In S. Batel & D. Rudolph (Eds.), *A critical approach to the social acceptance of renewable energy infrastructures: Going beyond green growth and sustainability* (pp. 161–178). Springer International Publishing.

Dartnell, L. R., & Kish, K. (2021). Do responses to the COVID-19 pandemic anticipate a long-lasting shift towards peer-to-peer production or degrowth? *Sustainable Production and Consumption, 27*, 2165–2177. https://doi.org/10.1016/j.spc.2021.05.018

Davies, S. E., & Wenham, C. (2020). Why the COVID-19 response needs international relations. *International Affairs, 96*(5), 1227–1251. https://doi.org/10.1093/ia/iiaa135

De Waal, F. (2006). ed. Stephen Macedo, Josiah Ober, with comments by Robert Wright, Christine M. Korsgaard, Philip Kitcher and Peter Singer. Primates and philosophers : Princeton University Press.

Dewey, J. (1922). *Human nature and conduct*. Courier Corporation.

Dewey, J. (1927). *The public and its problems: An essay in political inquiry*. Penn State Press.

Erdur, M. (2016). A moral argument against moral realism. *Ethical Theory and Moral Practice, 19*(3), 591–602. https://doi.org/10.1007/s10677-015-9676-3

Foucault, M. (1997). *The politics of truth*. Semiotext.

Galbraith, J. K. (1998). *The affluent society*. Hoghton Mifflin Company.

Geuss, R. (2001). *Public goods. Private goods*. Princeton University Press.

Gieryn, T. F. (1983). Boundary-work and the demarcation of science from non-science: Strains and interests in professional ideologies of scientists. *American Sociological Review, 48*(6), 781–795. https://doi.org/10.2307/2095325

Habermas, J. (1985). *The theory of communicative action: Volume 2: Lifeword and system: A critique of functionalist reason* (Vol. 2). Beacon press.

Habermas, J. (1996). Three normative models of democracy. In S. Benhabib (Ed.), *Democracy and difference. Contesting the boundaries of the political* (pp. 189–208). Pricneton University Press.

Harring, N., Jagers, S. C., & Löfgren, Å. (2021). COVID-19: Large-scale collective action, government intervention, and the importance of trust. *World Development, 138*, 105236. https://doi.org/10.1016/j.worlddev.2020.105236

Hartwell, C. A., & Devinney, T. (2021). Populism, political risk, and pandemics: The challenges of political leadership for business in a post-COVID world. *Journal of World Business, 56*(4), 101225. https://doi.org/10.1016/j.jwb.2021.101225

Heintz, J., Staab, S., & Turquet, L. (2021). Don't let another crisis go to waste: The COVID-19 pandemic and the imperative for a paradigm shift. *Feminist Economics, 27*(1–2), 470–485. https://doi.org/10.1080/13545701.2020.1867762

Hoy, D. C. (2005). *Critical resistance: From poststructuralism to post-critique*. MIT Press.

Jasanoff, S. (1990). *The fifth branch. Science advisers as policymakers*. Harvard University Press.

Kim, S. Y., & Grady, C. (2020). Ethics in the time of COVID: What remains the same and what is different. *Neurology, 94*(23), 1007–1008.

Korsgaard, C. M. (2015). Eternal values, evolving values, and the value of the self. *Foragers, Farmers, and Fossil Fuels: How Human Values Evolve, 2015*, 184–201.

Latour, B. (2013). Facing Gaia. Six lectures on the political theology of nature. *Gifford Lectures on Natural Religion, 2013*, 18–28.

Lindblom, C. E., & Cohen, D. K. (1979). *Usable knowledge: Social science and social problem solving*. Yale University Press.

Lowi, T. J. (1969). *The end of liberalism: Ideology, policy, and the crisis of public authority*. Norton.

Ludovic, J., Bourdin, S., Nadou, F., & Noiret, G. (2020). Economic globalization and the COVID-19 pandemic: Global spread and inequalities. *Bulletin of the World Health Organization, 2020*, 2–4.

Lynch, P., & Khoo, A. (2020). *Coronavirus: Volunteers flock to join community support groups*. BBC. In.

Macdonald, S., & Kam, J. (2007). Ring a ring o' roses: Quality journals and gamesmanship in management studies*. *Journal of Management Studies, 44*(4), 640–655. https://doi.org/10.1111/j.1467-6486.2007.00704.x

Mamo, L., & Fishman, J. R. (2013). Why justice? Introduction to the special issue on entanglements of science, ethics, and justice. *Science, Technology, and Human Values, 38*(2), 159–175. https://doi.org/10.1177/0162243912473162

March, R. J. (2021). Flatten the bureaucracy: Deregulation and COVID-19 testing. *The Independent Review, 25*(4), 521–536.

Mazzucato, M. (2018). *The value of everything: Making and taking in the global economy*. Hachette UK.

Merton, R. K. (1940). Bureaucratic structure and personality. *Social Forces, 18*(4), 560–568.

Michels, R. (2019). The iron law of oligarchy. In *Power in modern societies* (pp. 111–124). Routledge.

Moore, B. (1985). *Privacy. Studies in social and cultural history*. M.E. Sharp.

Murphy, J., Vallières, F., Bentall, R. P., Shevlin, M., McBride, O., Hartman, T. K., ... Hyland, P. (2021). Psychological characteristics associated with COVID-19 vaccine hesitancy and resistance in Ireland and the United Kingdom. *Nature Communications, 12*(1), 29. https://doi.org/10.1038/s41467-020-20226-9

Pesch, U. (2014). Sustainable development and institutional boundaries. *Journal of Integrative Environmental Sciences, 11*(1), 39–54. https://doi.org/10.1080/1943815X.2014.889718

Pesch, U. (2015). Publicness, privateness, and the management of pollution. *Ethics, Policy and Environment, 18*(1), 79–95. https://doi.org/10.1080/21550085.2015.1016960

Pesch, U. (2018). Paradigms and paradoxes: The futures of growth and degrowth. *International Journal of Sociology and Social Policy, 38*(11/12), 1133–1146. https://doi.org/10.1108/IJSSP-03-2018-0035

Pesch, U. (2020). Making sense of the self: An integrative framework for moral agency. *Journal for the Theory of Social Behaviour, 50*(1), 119–130. https://doi.org/10.1111/jtsb.12230

Pesch, U. (2020a). Making sense of the self: An integrative framework for moral agency. *Journal for the Theory of Social Behaviour, 50*(1), 119–130. https://doi.org/10.1111/jtsb.12230

Pesch, U. (2020b). A reply to "green shame: The next moral revolution?". *Global Discourse, 10*(2), 273–275. https://doi.org/10.1332/204378920X15785692888198

Pesch, U. (2021). Institutions of justice and intuitions of fairness: Contesting goods, rules and inequalities. *Critical Review of International Social and Political Philosophy, 2021*, 1–14. https://doi.org/10.1080/13698230.2021.1913887

Pesch, U., Huitema, D., & Hisschemöller, M. (2012). A boundary organization and its changing environment: The Netherlands environmental assessment agency MNP. *Environment and Planning C, 30*, 487–503.

Pesch, U., & Vermaas, P. E. (2020). The wickedness of Rittel and Webber's dilemmas. *Administration and Society, 52*(6), 960–979. https://doi.org/10.1177/0095399720934010

Rawls, J. (1997). The idea of public reason revisited. *The University of Chicago Law Review, 64*(3), 765–807.

Rawls, J. (2009). *A theory of justice.* Harvard University Press.

Rip, A. (1986). Controversies as informal technology assessment. *Science Communication, 8*(2), 349–371. https://doi.org/10.1177/107554708600800216

Roeser, S. (2005). Intuitionism, moral truth, and tolerance. *Journal of Value Inquiry, 39*(1), 75. https://doi.org/10.1007/s10790-006-3338-6

Rorty, R. (1989). *Contingency, irony, and solidarity.* Cambridge University Press.

Scharpf, F. W. (1997). *Games real actors play: Actor-centered institutionalism in policy research* (Vol. Vol. 1997). Westview Press Boulder, CO.

Sismondo, S. (2008). Science and technology studies and an engaged program. *The Handbook of Science and Technology Studies, 3,* 13–31.

Steinert, S. (2020). Corona and value change. The role of social media and emotional contagion. *Ethics and Information Technology, 23,* 59–68. https://doi.org/10.1007/s10676-020-09545-z

Taylor, C. (2002). Modern social imaginaries. *Public Culture, 14*(1), 91–124.

Tomasello, M., Carpenter, M., Call, J., Behne, T., & Moll, H. (2005). In search of the uniquely human. *Behavioral and Brain Sciences, 28*(5), 721–727. https://doi.org/10.1017/S0140525X05540123

Verweij, M., Van De Vathorst, S., Schermer, M., Willems, D., & De Vries, M. (2020). Ethical advice for an intensive care triage protocol in the COVID-19 pandemic: Lessons learned from the Netherlands. *Public Health Ethics, 13*(2), 157–165. https://doi.org/10.1093/phe/phaa027

Walby, S. (2021). The COVID pandemic and social theory: Social democracy and public health in the crisis. *European Journal of Social Theory, 24*(1), 22–43. https://doi.org/10.1177/1368431020970127

Warneken, F., & Tomasello, M. (2006). Altruistic helping in human infants and young chimpanzees. *Science, 311*(5765), 1301–1303. https://doi.org/10.1126/science.1121448

Weintraub, J. (1997). The theory and politics of the public/private distinction. In J. Weintraub & K. Kumar (Eds.), *Public and private in thought and practice. Perspectives on a grand dichotomy* (pp. 1–42). University of Chicago Press.

Open Access This chapter is licensed under the terms of the Creative Commons Attribution 4.0 International License (http://creativecommons.org/licenses/by/4.0/), which permits use, sharing, adaptation, distribution and reproduction in any medium or format, as long as you give appropriate credit to the original author(s) and the source, provide a link to the Creative Commons license and indicate if changes were made.

The images or other third party material in this chapter are included in the chapter's Creative Commons license, unless indicated otherwise in a credit line to the material. If material is not included in the chapter's Creative Commons license and your intended use is not permitted by statutory regulation or exceeds the permitted use, you will need to obtain permission directly from the copyright holder.

Part II
Envisioning a Post-Pandemic Future

Chapter 8
Offsetting Present Risks, Preempting Future Harms, and the Ethics of a 'New Normal'

Sven Nyholm and Kritika Maheshwari

8.1 Introduction

Since the onset of the COVID-19 pandemic, we have witnessed several changes and shifts in our ordinary, everyday behaviour. In an effort to mitigate and tackle the risk of global spreading of the SARS-CoV-2 virus, governments, scientists, and various health organisations proposed adopting new safety precautions. For instance, working and schooling small children from home became the norm amongst those who had the option to do so. In addition, wearing a facemask to the supermarket or keeping a 1.5 m distance from one another when in public spaces suddenly became widespread in many parts of the world.

For most people, the idea of undertaking precautions like wearing a medical facemask unless one is in a hospital or of not being permitted to visit restaurants nor work in their workplace would have seemed incredible before. But suddenly, this was, and in some places continues to be, what many referred to as "the new normal". For instance, in implementing mandatory face masks as a national health policy, India's Union Government notified its citizens, saying, "we should

S. Nyholm (✉)
Utrecht University, Utrecht, Netherlands
e-mail: s.r.nyholm@uu.nl

K. Maheshwari
University of Groningen, Groningen, Netherlands
e-mail: k.maheshwari@rug.nl

© The Author(s) 2022
M. J. Dennis et al. (eds.), *Values for a Post-Pandemic Future*, Philosophy
of Engineering and Technology 40, https://doi.org/10.1007/978-3-031-08424-9_8

incorporate it in our lives as a new normal".[1] Similarly, in announcing the return of mandatory masks in public as a long-lasting social practice, the Governor of Connecticut, Ned Lamont, informed his fellow Americans that, "We are going to be getting back to normal; it will be a new normal".[2]

While discussions of a new normal gained momentum during the COVID-19 pandemic, what exactly people mean when they speak and write about it is not always clear. Moreover, it is noteworthy that there are also various other cases that can be and that have sometimes been classified as being a new normal: for example, this can happen when new technologies are introduced (e.g. a new normal emerges in a world in which self-driving cars populate public streets) or when other changes occur (e.g. a new normal emerges in a world with drastic climate change). While there is an extensive discussion of pandemic ethics, and also independent discussions of related cases in the literature, the idea of exploring a range of different cases under the umbrella of "a new normal" at a general level remains nascent and under-explored.

Our aim in this chapter is to engage in preliminary groundwork for what we will call the "ethics of a new normal" more generally. We are interested in what different kinds of situations that can be viewed as involving a new normal have in common from an ethical point of view, and we will identify a number of key considerations that are likely to be relevant in most instances of what could be called the ethics of a new normal. We think it is useful not only to think about those different kinds of cases in isolation, but also in relation to each other, with an eye to what they have in common. This can be useful for engaging with or discussing a new normal that is already a reality – like the COVID-19 pandemic – or a new normal that we are already transitioning towards. It can also be useful when we discuss a new normal that we are likely to transition to – like a future with driverless cars or a future dealing with the problems associated with climate change.

Our discussion proceeds as follows. In Sect. 8.2, we offer a schematic definition of what we mean by "the ethics of a new normal" and explicate its relationship with familiar topics in the literature. Section 8.3 explores different examples of concrete ethical discussions that can be classified as instances of the ethics of a new normal. Section 8.4 identifies some morally relevant distinctions important for discussing the ethics of a new normal. This section provides a classification of such distinctions, which is summarized in a table at the end of the section. In Sect. 8.5, we discuss shortcomings of popular hardliner arguments for transitioning towards the new normal offered in the literature. Finally, we propose an alternative: drawing upon John Broome's discussion of offsetting climate harms, we discuss a general risk-offsetting principle as a plausible alternative for thinking about how to respond to a new normal. Section 8.6 concludes.

---

[1] https://www.livemint.com/news/india/face-masks-are-new-normal-incorporate-it-in-our-lives-centre-amid-third-covid-wave-fears-11626432848811.html (Accessed on September 19, 2021)

[2] https://www.rnz.co.nz/news/world/414437/covid-19-worldwide-cases-exceed-2-million-death-toll-crosses-136-600 (Accessed on September 19, 2021)

8.2   Outlining the "Ethics of a New Normal"

Addressing any ethical or political issues surrounding any so-called new normal must presuppose an account of what a new normal is or what it represents. For instance, in recent times of COVID-19, academic and public discussions mainly revolved around questions of whether and how our "new normal" ways of responding to pandemic risks should or could be our "new future" (e.g. Bramble, 2020). Little attention, however, is paid to the very idea of the "new normal" itself. What, for instance, makes a situation or a particular state of affairs a "new normal" in a morally and/or politically relevant sense? Or, to put it differently, when and why do certain change(s) in affairs indicate the advent or beginnings of an actual or potential new normal that deserves our attention?

While there is no definitive understanding of the notion of "a new normal", it appears to be common to use this phrase when some new risk has been introduced and new safety precautions are therefore called for. Jeff Clyde Corpuz (2021) notes that the first use of "a new normal" was likely during the time of the 2008 financial crisis. Back then, the term referred to the various impacts the crisis had on individuals and society at large, such as conditions of precariousness and long periods of social unrest. In current discussions of a new normal, many people seem to refer to the idea of a shift, change, or transition in an existing or established set of individual or structural risk-taking or risk-creating practice(s) or behaviour(s).

We will mostly be focusing on issues related to likely harms and other risks below, since that seems to be a common factor in many discussions of a new normal. But on the most general level, we will take it that "a new normal" refers to a situation or a state of affairs that is different relative to some temporal baseline. The difference in the state of affairs may track changes or shifts in our social, moral, legal, economic, epistemic, psychological, biological behaviour, practices, norms, or a combination of these. In each case it would be a "new normal" in light of one factor or set of factors that has either been introduced, removed, or modified in a way that impacts how we live our day-to-day lives, conduct our current affairs, or what we consider as normal. By "the ethics of a new normal", we mean any and all ethical questions having specifically to do with such factors that contribute to the existence of a new normal.

What we are calling the "ethics of the new normal" here is not often discussed in these general terms. But it is closely related to, and partly overlaps with, various other ongoing and well-established discussions within applied ethics and moral philosophy more generally. Here are some examples that we can briefly consider in order to situate what we are calling the ethics of a new normal in relation to other on-going debates. (In the next section, we will consider concrete examples of ongoing debates that we think are clear instances of the ethics of a new normal.)

First, there is the ethics of risk more generally (E.g., Hansson, 2003; Hayenhjelm & Wolff, 2012; Maheshwari, 2021). This is the discussion of how we should deal with risks and uncertainty in life, why it is bad to be exposed to risk, what is wrong with imposing risks on others, how much risk is acceptable in life, what safety precautions are ethically required in different parts of life, and so on. Such questions about the ethics of risk are highly relevant in any discussion of the ethics of a new normal. The reason for that is that a new normal can often be understood as a situation where one or more new risks or sources of uncertainty have been introduced into human life, which makes a difference to how we have reason to go about our everyday lives.

Second, there is the ethics of resilience (Cañizares et al., 2021). This is a set of ethical issues related to how resilience can or should be achieved in essential parts of human life. This is often discussed within engineering ethics. There is an important paradigm within engineering research, "resilience engineering", which is about how the outputs of engineering projects (bridges, socio-technological systems, and so on) can be made resilient against different kinds of pressures (Doorn, 2021). Resilience is also an important concept within discussions about health and how to protect the health of human beings and non-human animals, or perhaps even the health of whole ecosystems. The ethics of resilience is highly relevant to, but also partly different from, what we call the ethics of a new normal. Resilience is usually about maintaining something that already exists or maintaining something that is being created. In the ethics of a new normal, the main issue is dealing with new challenges that affect our day-to-day lives. Thus, matters related to resilience are central to the ethics of a new normal. But other questions are paramount as well: for example, whether to respond to new challenges to our everyday lives by creating new forms of safety precautions or design other forms of harm- or risk reduction strategies.

Third, one more thing – the last thing we will mention here – that comes to mind as partly overlapping with and certainly relevant for the ethics of a new normal is the debate about transitional justice (E.g. Teitel, 2000; Murphy, 2012). When academics and others discuss transitional justice, they usually talk about protecting human rights after some armed conflict, natural disaster, or other events that might raise questions about justice after a transition from one state of affairs to another. Such discussions can be seen as being about a new normal. Accordingly, we view transitional justice debates as being included in the general category of the ethics of a new normal. But we understand the ethics of a new normal to cover a wider range of issues than those very important ones that specifically have to do with human rights and justice that are part of transitional justice debates. We will now put forward a number of other examples of concrete topics and debates that we view as important instances of the ethics of a new normal.

8.3   Examples of Ethical Debates that Can Be Viewed as Part of the Ethics of a New Normal

As noted above, our characterisation of a new normal is intended to be general and expansive to the effect that it allows us to situate various ongoing ethical debates about current, future, or potential new normal scenarios under the broader umbrella of "the ethics of a new normal". In this section we identify a number of discussions and specific contributions to practical ethics that can be viewed as being part of the ethics of a new normal. As we see things, it is often useful to compare ethical discussions that are going on in parallel, that might have interesting similarities, and that can therefore be interesting to discuss side-by-side under a general heading. Hence we see it as useful to consider whether different ongoing discussions in practical ethics can be viewed as fitting under the heading of the ethics of a new normal. Additionally, we think that there are cases where the term "a new normal" might not have been used explicitly but where it nevertheless makes sense to compare those cases with other cases that have explicitly been labeled as a new normal.

This holds, for instance, in the case of discussions of disruptive technologies that are on the horizon but are not yet widespread in society. Take the introduction of self-driving cars or artificially intelligent humanoid robots that may be on the horizon (Nyholm, 2020; Royakkers and van Est, 2015). These can also be seen as bringing about a new situation where there is a new key factor or set of factors that people need to consider when going about their day-to-day lives. When it comes to humanoid robots, philosophers have begun discussing whether a new normal involving such robots would require us to give certain rights to those robots or treat them with some degree of moral consideration. (Nyholm, 2020: Chap. 8) Consider next self-driving cars. In recent years, there has been a lot of hype about the large-scale introduction of autonomous vehicles, also known as self-driving cars, into public traffic (Gurney, 2016). Both academic and public discussions are filled with speculation and anticipation of how things could or should be in light of changes in our routine driving practices if and when self-driving cars on our public roads becomes the new normal (e.g. Royakkers and van Est, 2015).

For instance, would we no longer need to avoid drinking alcohol before using a car? Would driving as a profession no longer exist? Or – to also bring up a much-discussed topic – how should these self-driving cars handle risky scenarios where accidents appear to be unavoidable? And who should be held responsible if and when a self-driving car injures or kills a human being? (Nyholm, 2018a; b). Being faced with such questions will soon, many think, not only be a hypothetical future scenario. It will instead be a new normal within the domain of traffic (Gogoll & Müller, 2020). The same can be said about many other technologies that are currently primarily part of visions of what the future will be like: e.g. our other just-mentioned example of maximally humanlike robots with high degrees of artificial intelligence (e.g. Danaher, 2020; Schwitzgebel & Garza, 2020). We are not yet living in a world featuring such technologies. But before we know it, it might be the new normal.

Likewise, the developments related to climate change and the factors differentiating the modern world from previous times in human history are examples of new factors providing reasons to reflect on how we live our lives (Jamieson, 2014; Di Paola, 2017). Frequent very extreme weather conditions – such as hurricanes or tornados – and much higher average temperatures, and all of the impacts this will have on nature, are often discussed as things that future people will have to deal with as part of their everyday experience. It might soon become the new normal that temperatures are sweltering, that extreme weather changes become much more common, and that an increased number of places in the world will become uninhabitable for human beings and many non-human animals (Broome, 2012). This is another expected transition to a new normal that raises numerous pressing ethical questions about the ethical defensibility of our current ways of living and consumption patterns.

In two relatively recent books that are worth mentioning in this context – *Death and the Afterlife* (2011) and *Why Worry about Future Generations* (2018) – Samuel Scheffler's discussion of the relationship between current and future generations appears to fall under the ethics of a new normal. According to Scheffler, it matters to us more than we might realise whether or not there will be people around after we are gone. To illustrate this, one of the things Scheffler does is to ask his readers to imagine the prospects of Armageddon-like scenarios, wherein we realise that we are the last generation of people who will ever live (also c.f. Ord, 2020). The scenarios Scheffler discusses are inspired by science fiction. But his aim in using such scenarios is to try to get us to realise what matters to us in real life when it comes to our relation to the people who will exist – or perhaps not exist, depending on how things go – in the future, after we ourselves are dead.

In one scenario Scheffler lays out, a giant asteroid is travelling through space towards the planet Earth. It is clear that the collision between this asteroid and our planet will cause so much damage that human life on Earth will no longer be possible after the collision. Thus, the people in Scheffler's example face the "new normal" of realising that they are the last people who will live and that human life will end soon.

In a second scenario – inspired by P.G. James's novel *The Children of Men* – there is no asteroid travelling towards Earth that is about to suddenly kill everyone in a violent collision. However, human life will end for another reason. Everyone has become infertile, and nobody has conceived a child for the last 25 years. This universal infertility is not the result of anything anybody did. Moreover, people are otherwise healthy, and it seems that they will get to live out their full lifespan. But nobody will come after them since nobody is having any children. This is another "new normal" that Scheffler asks us to consider and react to (though he does not use the expression "new normal").

Scheffler's aim in putting these examples forward and asking his readers to react to them is to try to make us see what value we put – either explicitly or at least implicitly – on there being people who will live after we are gone, and who can carry on our projects, and who can value the sorts of things we value. The normal situation is that we safely assume that there will be people coming after us. But with climate change and other existential threats to humanity, the new normal might become or already be that we cannot anymore simply take for granted that future generations will take over after we are gone. (Scheffler, 2011, 2018, see also Ord, 2020 and Nyholm, 2021).

Another striking instance of the ethics of a new normal appears in the book *Unfit for the Future,* wherein Ingmar Persson and Julian Savulescu (Persson & Savulescu, 2012) argue that we today face risks – including risks of what they call "ultimate harm", viz. potentially irreparable damage to the possibility of future human life on Earth – that our evolved human psychology has not been adapted to help us deal with. In comparing the circumstances in which human beings have lived during most of the history of our species with the circumstances of living in the modern world, Persson and Savulescu, in effect, argue that we seem to be part of a new normal whereby the modern world involves ethical challenges – related to existential risks and large-scale collective action problems – that our evolved human psychology is not ready for. Thus, we are, Persson and Savulescu argue, "unfit for the future" to an extent that raises urgent ethical questions about how to deal with or respond to these "new" aspects of the new normal.

We are bringing up this example – not only to give yet another example of something that could be classified as a case of the ethics of a new normal – but also to illustrate what counts as a new normal is a relative matter. Compared to the 150,000–200,000 years of the history of our species, life in the modern world is a unique situation for our human species. Most of human history so far took place in a very different kind of world, as described by Persson and Savulescu in their discussion. That is an extremely long-term perspective.

That situation described from this long-term perspective is one kind of new normal. It is a new normal in the grand scheme of things. But, by a new normal, we might also mean something much more abrupt than the transition from prehistoric times to the present. We might mean the transition between, say, everyday life as it was in 2019 before the COVID-19 pandemic and life as it came to be, in most of the world, thereafter, during, say, 2020 and 2021. For instance, the COVID-19 pandemic has recently introduced new factors that we have reason to consider when we decide how to live our everyday lives, namely, the presence of the Coronavirus and its dangers and risks (Bramble, 2020). The switch to wearing masks or keeping a 1.5 meter distance to reduce one's potential exposure to risk, for instance, marked an instantaneous change in our social behaviour and practice.

As we have just seen, there is a range of different discussions that can be seen as all being part of the overall topic of the ethics of a new normal. We are interested in what different cases of an ethics of a new normal have in common. And we are here looking at these kinds of cases from a zoomed-out perspective to see whether there are any general distinctions, considerations, or principles that are likely to be relevant in most or all of these cases. What we are calling the "ethics of the new normal" here is not often discussed in these general terms, as far as we know. But as we have just seen, various ongoing and well-established discussions within applied ethics and moral philosophy more generally are clear instances where what is being the discussed is the ethics of a new normal.

8.4   Some Key Distinctions

To analyse any given situation that is a new normal or likely to become the new normal, further clarification regarding the scope, the depth, and the breadth of the ethical questions that may arise within its domain is required. The exact details of different situations that can be labeled a new normal will differ, but it is also possible to reflect on ethically relevant features that many, if not most, situations that can be labeled a new normal might have in common. To this end, we will now note some basic, general distinctions that are likely to be relevant in most or all of these cases.[3]

First, we distinguish whether everyone currently existing or who will exist, or only some subset of this population is affected by a new normal. This distinction has two applications: first, we can ask who has reason to change their ways of going about their everyday lives because of the new normal. And second, we can ask whose lives are directly impacted by any good or harmful effects of the new normal. Sometimes, the answers to those questions can refer to one and the same group. But there might only be a partial overlap. Or perhaps they are two separate groups.

Consider climate change as an example. Those who live now and might be concerned about how their lifestyles might adversely impact the climate may feel that they have reason to change their day-to-day behaviour to lessen the harmful effects climate change might have on the lives of people who will live after we are all dead. Or we might have a case where the people affected by the new normal also have reason to change their normal habits. This happened for most people in the world during the COVID-19 pandemic. Everyone faced the risk of getting the virus, and everyone, or almost everyone, had reason to change their day-to-day habits.

With climate change, by contrast, it might first be the case that only some are primarily affected. For example, for the people living in the Maldives, climate change will soon have an extreme impact since the island nation will most likely end up completely underwater within 75 years or so.[4] Later on, when climate change becomes more extreme, it might dramatically affect all people worldwide.

Relatedly, we can draw a distinction between whether life, in general, is affected by the new normal or whether there is primarily one part of life that might be affected by a new normal. So, for example, with something like climate change or a pandemic, it may be that life in general – or almost every aspect of life – is impacted by the new normal. But when it comes to introducing some new technology, it may be that it is primarily one specific part of life that is affected.

Of course, there might be ripple or cascading effects and repercussions on other aspects of life, but the central part of life where there is a new normal might be some clearly definable part of life. If fully self-driving cars are introduced as an option, we can use them on the road, for example, and the part of life that might be regarded as a new normal is primarily the domain of traffic. This might be a case where there

---

[3] Note that this list of distinctions is not exhaustive.

[4] https://www.worldbank.org/en/news/feature/2010/04/06/climate-change-in-the-maldives (Accessed on September 19, 2021)

are various further effects and repercussions – if we are to believe some of the speculations about what will happen when fully self-driving cars become widespread! – but it is most obviously traffic where there is a new normal (Royakkers and van Est, 2015).

Another distinction that is relevant to draw is between whether the new normal is likely to be a transitory phase or whether the new normal is something more permanent. For example, a pandemic might mean a new normal where life is turned upside-down for a while – perhaps even a long time. But in the end, things might "go back to normal" in the sense that they will resemble how things were before the pandemic again. In contrast, when climate change becomes extreme, it might be that there is no turning back; the new normal might be here to stay. This overlaps with the distinction between whether some new normal is reversible or whether it is irreversible.

For instance, in case of an ongoing pandemic, unless a vaccine is invented with an extremely high efficacy or other ways of dealing with a pandemic are worked out, life in a world with a novel virus might irreversibly be a new normal, where we cannot go back to how things were before. But with the right technologies and other ways of tackling the multiple problems of this pandemic, it might be that the new normal is reversible and that there is a way of going back to how things were before. With something like climate change caused by large-scale pollution and resource depletion, in contrast, the problem might be irreversible, and the new normal might be there to stay. It should be noted, though, that what is apparently irreversible might not be de facto irreversible. Even something like climate change might, according to some technology optimists, be a reversible problem.[5]

This points to a closely related distinction of whether a new normal results from human agency or some natural cause outside of human control. In other words, is it anybody's fault – our own fault perhaps – that something has become the new normal or might soon become the new normal? Was it because of something that some individual or group of individuals – or some big collective of people – did? Or did the new normal come about because of some other reason? For example, during the COVID-19 pandemic, there was widespread speculation over whether the virus was naturally occurring or was created in a laboratory by human beings. Similarly, whether climate change depends on our human lifestyles or primarily on other causes is something that animates those concerned about climate change.

In general, whether a new normal is the result of human agency or not matters greatly to what ethical categories of assessment are appropriate to use in the evaluation of options and when we think about how it is appropriate to respond to the new normal. So, for example, should we think in terms of whether people living now have duties of goodness or duties of justice towards future generations, to use two expressions from John Broome's (2012) book *Climate Matters: Ethics in a Warming World*? Or should we instead think in Shefflerian terms of whether we have reasons

---

[5] See, for instance, the discussion about geo-engineering in Scott (2012).

Table 8.1 Key distinctions and concepts within the ethics of a new normal

New normal (Examples)	For some		For everyone	
	In limited domains of life:	In all aspects of life:	In limited domains of life:	In all aspects of life:
	For example: People who drive cars in a new normal that involves having the option of using fully self-driving cars	For example: Climate change in the short-run (e.g., for people living in the Maldives)	For example: What is involved in appearing in specific public spaces during a pandemic	For example: A pandemic is so severe that new safety precautions have to be taken in all parts of life
Reversibility?	The new normal could involve irreversible damage to some or all populated regions. Example: Climate-change induced flooding, extinction risk (irreversible for all)			
Repeatability?	The new normal could either be a one-off incident or a repeatable one.			
Reparability?	The new normal could or could not involve harm that can be redressed, or offered restitution for, to all or some affected particles.			
Responsibility?	The new normal could be due to an individual's or a group's fault or responsibility.			

to feel some form of despair, resignation, or existential sense of meaninglessness when we think of worrying aspects of a new normal? (Nyholm, 2021).

There are four broad classes of cases that seem to call for different forms of normative assessment. In one kind of case, the new normal is nobody's fault, and it might be unclear whether there is anything we can do about it. By contrast, secondly, there can also be cases in which the new normal is nobody's fault, but where there is something that can be done. Thirdly, there are cases in which the new normal is somebody's fault and where something can be done about it. Fourth and lastly, there is, of course, also the grimmer type of case where the new normal is somebody's fault – either some individual or some group's fault – but where there is nothing that anybody can do to deal with whatever problems or risks are involved in the new normal. These four kinds of scenarios differ with respect to what normative duties or appropriate responses we should associate them with.

We can summarize some of the above-considered distinctions in a matrix as follows (Table 8.1):

8.5 The Hardliner Vs the Offsetting Approach to the Ethics of a New Normal

As noted above, in his earlier-mentioned book, Broome (2012) distinguishes between duties of goodness (or duties of benevolence), on the one hand, and duties of justice, on the other hand. If, for example, there is something we can do to deal with some problem associated with a new normal, but it is nobody's fault that there is this new normal (e.g. there is a naturally occurring virus), it might be that our duties to act are not associated with rectifying some injustice but that our ethical duties are instead related to what it would be good or benevolent to do. By contrast, if some problem (e.g. the issues related to human-created climate change) is our fault, we might have what Broome calls duties of justice to change the way we live our lives to try to counteract the problem. Lastly, suppose a problem associated with a new normal is our fault, but that nothing can be done about it. (According to some people's estimates, this might soon be the case in relation to human-created climate change. (Persson & Savulescu, 2012)) In that case, the normatively appropriate response might be deep regret or a realisation that what we have done is the opposite of something positively meaningful (Nyholm, 2021).

In most cases, the new factor or set of factors that creates a new normal is some risk or set of risks. Or it might be some new technology that can be used to deal with risks in a novel way. In each type of case, risks and safety considerations are central to what helps to define something as a new normal. For example, the new normal related to the COVID-19 pandemic had to do with the new risks introduced by the Coronavirus. Similarly, the new normal pertaining to climate change has to do with new risks of great harm related to living in a different type of climate. And, to give one more example, new technologies (such as fully self-driving cars) might be claimed to have as their primary benefit to make some activity (in this case, driving) safer (Gurney, 2016). So, the duties of goodness and duties of justice related to a new normal are typically going to be duties of goodness or justice of how to respond to risks. The last thing we will do is, therefore, to briefly discuss the ethics of how to deal with risks and risk management concerning a new normal.

Notably, when a new normal comes about or is about to come about, and this involves introducing new risks or new ways of mitigating risks, this will often lead some to suggest what we will call a hardline approach. For example, in a recent article about the COVID-19 pandemic, Peter Singer (2021) argues that there is an obligation that applies to everyone that we should all get vaccinated. This should not, as some think, be a matter of personal choice.[6] In discussing this issue, Singer makes a comparison between compulsory COVID-19 vaccinations and compulsory seat-belt use. We should have the former, just like we should have the latter, and for similar reasons, Singer argues. Call this a hardliner approach.

This way of arguing – i.e., approaching the ethics of risks by comparing one type of safety precaution with another – is common in the specific discussions we have identified above as being instances of the ethics of a new normal (cf. Giubilini & Savulescu, 2019). It is common in the ethics of dealing with the COVID-19

---

[6] A member of the United States archery team that competed in the Tokyo 2020 Olympics – Brady Ellison – took that view, decided not to get vaccinated, and claimed that such a decision is "one hundred percent a personal choice," and that "anyone that says otherwise is taking away people's freedoms." In arguing for compulsory vaccinations, Singer was reacting to such views – even directly quoting the athlete Ellison disapprovingly – and he argues that a view like Ellison's falsely makes it appear as if the only person affected by their choice not to get vaccinated is the person him or herself.

pandemic, impending climate change, and the introduction of new technologies, such as fully self-driving cars.

For example, Jan Gogoll and Julian Müller (Gogoll & Müller, 2020) discuss the choice between fully self-driving cars and regular cars in a new normal in which we would face such a choice. They argue that if self-driving cars would be safer than standard cars, then everyone should be made to use self-driving vehicles rather than regular cars. To make this case, they appeal to an argument that Jason Brennan (2018) has put forward in the context of vaccination ethics for why, according to him, even libertarians can be in favour of compulsory vaccinations. The general idea, borrowed from Sven Ove Hansson, is that imposing risks on others is only permissible if it is part of a practice that works to everyone's benefit and there is no alternative practice that works better to everyone's benefit (Hansson, 2003).

In this hardliner picture, something like the practice of mask-wearing, for instance, as a way to reduce one's chances of imposing risks on others involves mandatory transitioning from an "old normal", where wearing masks was not part of a risk-mitigating or risk-reducing socially beneficial activity, to a "new normal", where this is the case. The introduction of autonomous self-driving cars deemed safer than ordinary cars may involve a similar type of transition. It might involve transitioning from the "old normal", where driving ordinary cars is considered a socially beneficial activity that involves mutually advantageous risk-taking for members of a society, to a "new normal", where this is the case only for autonomous self-driving cars.

These new ways may entirely override, replace, or supplant existing safety norms and practices of taking precautionary and preventative actions (or omissions). If we consider such transitions only with an eye to the aim of promoting safety, and we set other considerations aside, moving to the safer alternative will often seem like the right thing to do. However, this idea of mandatorily transitioning into a new normal by changing one's behaviour in response to new risks is not without controversies, for it overlooks certain feasibility considerations. For instance, in many places, people either simply failed to afford taking precautionary measures like wearing expensive masks because of lack of financial resources or failed to observe recommended actions like maintaining distance in public spaces was simply not physically feasible due to shortage of space.

What would be another approach? One type of approach we have in effect already considered: namely, the idea that it should be a personal choice whether one wants to take some form of safety precaution in response to the risks associated with some new type of situation. This was the approach that the Olympic archer Brady Ellison advocated concerning COVID-19 vaccines, as mentioned in footnote 6 (Singer, 2021). It is also an approach that some have voiced their approval of when it comes to choosing between fully self-driving cars and manually driven cars.

Something called the "human driving association" has published a "human driving manifesto", in which they argue that people's freedom to drive regular cars should not be taken away if fully self-driving cars are introduced into society (Roy, 2018). This organisation is, they say, "pro steering wheel"; they do not want to be forced to drive self-driving cars, even if they would turn out to be safer than regular cars. Interestingly, however, the rest of the "human driving manifesto" signals a willingness to advocate extra safety precautions for those who wish to continue driving regular cars even as the option of what would supposedly be safer self-driving cars is introduced. Thus, for example, the members of the human driving association signal a willingness to make the tests one has to pass to get a driver's license much more demanding so that only those who drive in a very safe way can get a license.

They also show openness to other forms of technological safety precautions, such as lane-keeping assistance technologies. And one can also imagine things such as alcohol locks and speed regulation technologies that could help make manual driving safer than it might be at present. These would be ways of compensating for the additional risks associated with regular cars to make the safety level of manual driving more like that of self-driving cars (Nyholm & Smids, 2020).

This is reminiscent of an approach that Broome (2012) suggests regarding alternative ways of dealing with our carbon footprint and climate change. Broome does not argue against the hardliner approach of doing things that would lessen one's carbon footprint, e.g., no longer travelling in aeroplanes, driving less, or whatever. But he does argue that there is an equally defensible alternative. Broome argues that we can compensate for or "offset" our carbon footprint and that this can be equally acceptable to minimising our carbon footprint.

How could one offset one's carbon footprint? Broome discusses options such as planting trees, paying some form of climate tax that could be used to compensate people of the future, or – and this is an exciting idea! – buying environmentally friendly stoves for people in communities where they cook with environmentally unfriendly stoves. If we do these kinds of things that help to neutralise our impact – if we put back as much as we take away – this is equally good as trying to lessen or minimise our carbon footprint, Broome argues.

A more general principle along these lines could be something like the following: when a new normal comes about, and there is some new option that helps to counteract the risks associated with the new normal, we should either go for this option or, alternatively, make use of other safety precautions that could help to offset the risks we create by not making use of that seemingly safest option. Call this the risk-offsetting principle (Cf. Nyholm, in press).

Suppose somebody does not wish to be vaccinated against COVID-19, or that they do not want to wear a facemask even though that is deemed to be an efficient safety precaution. In theory, they could potentially get away with this from an ethical point of view if they took some other form of safety precaution. For example, they could get away with it if they consistently maintained a safe distance (over one and a half meters) between themselves and other people. Of course, in practice, this will be hard to do for most people, but in theory it could be a way of offsetting the risks for others that are created by not getting vaccinated or not wearing a facemask.

Or suppose self-driving cars indeed become safer than regular cars, but somebody wants to continue driving a manually driven car. In that case, they could potentially get away with this from an ethical point of view if they took added safety precautions (alcohol locks, speed limiting technologies, lane-keeping assistance technologies, or whatever) that would help to offset the risks associated with driving a regular car rather than using a self-driving car. These kinds of choices – either to use the newer, safer option or to use additional safety precautions when going for some other, less safe option – are likely to arise in many situations involving a new normal when some new risk or some new form of safety precaution is introduced into some part of life. Accordingly, we think that a general principle such as the one just sketched is likely to be relevant in most cases we are dealing with the ethics of a new normal.

8.6 Concluding Remarks

As noted above, when the COVID-19 pandemic was at its height, and people were forced to take various forms of safety precautions in their day-to-day lives that previously could not have been imagined (wearing facemasks, working from home, not going to bars and restaurants, and so on), some started speaking about a "new normal". As we have noted above, ethical discussions about how we might need to change our behaviour in response to a new situation are not limited to the case of a pandemic. Other forms of development can also create a new normal, either concerning life in general or only in some specific domain. And similar ethical questions about how we should change our everyday behaviour arise then as well. For example, climate change is increasingly putting pressure on whether our day-to-day ways of living in the modern world are acceptable. Or if a new technology (e.g., fully self-driving cars) is introduced into society and functions as a game-changer in terms of risks and safety precautions, this is also part of the ethics of a new normal.

In this chapter, we have zoomed out to a general level and considered the very idea of a new normal and reflected on what broad ethical distinctions, considerations, and principles might be worth considering in most cases where we are facing an ethics of a new normal. Our discussion above is by no means a complete account of how we should approach the ethics of a new normal. It is a sketch of some of the considerations, distinctions, and principles that seem like they will be relevant in most discussions of this sort. More work is needed. We hope that our remarks above might stimulate others to zoom out to the more general level and ask what different cases of a new normal might have in common from an ethical point of view.[7]

---

[7] We are grateful to an anonymous reviewer and to the editors of this volume. Sven Nyholm's work on this article is part of the research program Ethics of Socially Disruptive Technologies, which is funded through the Gravitation program of the Dutch Ministry of Education, Culture, and Science and the Netherlands Organization for Scientific Research (NWO grant number 024.004.031).

References

Bramble, B. (2020). *Pandemic ethics*. Bartleby Books.

Brennan, J. (2018). A libertarian case for mandatory vaccination. *Journal of Medical Ethics, 44*(1), 37–43. https://doi.org/10.1136/medethics-2016-103486

Broome, J. (2012). *Climate matters: Ethics in a warming world*. Norton Books.

Cañizares, J. C., Copeland, S., & Doorn, N. (2021). Making sense of resilience. *Sustainability, 13*(15), 8538. https://doi.org/10.3390/su13158538

Corpuz, Jeff Clyde G. (2021). Adapting to the culture of a 'new Normal': An emerging response to COVID-19. *Journal of Public Health, 43*(2), e344–e345. https://doi.org/10.1093/pubmed/fdab057

Danaher, J. (2020). Welcoming robots into the moral circle: In defence of ethical behaviourism. *Science and Engineering Ethics, 26*(4), 2023–2049. https://doi.org/10.1007/s11948-019-00119-x

Di Paola, M. (2017). *Ethics and politics of the built environment: Gardens of the Anthropocene*. Springer.

Doorn, N. (2021). The role of resilience in engineering. In D. Michelfelder & N. Doorn (Eds.), *The Routledge handbook of the philosophy of engineering* (pp. 482–493). Routledge.

Giubilini, A., & Savulescu, J. (2019). Vaccination, risks, and freedom: The Seat Belt analogy. *Public Health Ethics, 12*(3), 237–249. https://doi.org/10.1093/phe/phz014

Gogoll, J., & Müller, J. (2020). Should manual driving be (eventually) outlawed? *Science and Engineering Ethics, 26*(3), 1549–1567. https://doi.org/10.1007/s11948-020-00190-9

Gurney, J. K. (2016). Crashing into the unknown: An examination of crash-optimization algorithms through the two lanes of ethics and law. *Albany Law Review, 79*(1), 183–267.

Hansson, S. O. (2003). Ethical criteria of risk acceptance. *Erkenntnis, 59*(3), 291–309. https://doi.org/10.1023/A:1026005915919

Hayenhjelm, M., & Wolff, J. (2012). The moral problem of risk impositions. *European Journal of Philosophy, 20*(1), E1–E142. https://doi.org/10.1111/j.1468-0378.2011.00482.x

Jamieson, D. (2014). *Reason in dark times: Why the struggle against climate change failed – And what it means for our future*. Oxford University Press.

Maheshwari. (2021). On the harm of imposing risk of harm. *Ethical Theory and Moral Practice, 24*, 965–980. https://doi.org/10.1007/s10677-021-10227-y

Murphy, C. (2012). *The conceptual foundations of transitional justice*. Oxford.

Nyholm, S. (2018a). The ethics of crashes with self-driving cars: A roadmap, I. *Philosophy Compass, 13*(7), e12507. https://doi.org/10.1111/phc3.12507

Nyholm, S. (2018b). The ethics of crashes with self-driving cars: A roadmap, II. *Philosophy Compass, 13*(7), e12506. https://doi.org/10.1111/phc3.12506

Nyholm, S. (2020). *Humans and robots: Ethics, agency, and anthropomorphism*. Rowman and Littlefield International.

Nyholm, S. (2021). Meaning and anti-meaning in life and what happens after we die. *Royal Supplements in Philosophy, 90*.

Nyholm, S. (in press). The ethics of transitioning towards a driverless future: Traffic risks and the choice among cars with different levels of automation. In D. Michelfelder (Ed.), *Test-driving the future* .Rowman and Littlefield International.

Nyholm, S., & Smids, J. (2020). Automated cars meet human drivers: Responsible human-robot coordination and the ethics of mixed traffic. *Ethics and Information Technology, 22*(4), 335–344. https://doi.org/10.1007/s10676-018-9445-9

Ord, T. (2020). *The precipice: Existential risk and the future of humanity*. Hachette Books.

Persson, I., & Savulescu, J. (2012). *Unfit for the future*. Oxford University Press.

Roy, A. (2018). This is the human driving manifesto: Driving is a privilege, not a right. Let's fight to protect it. *The Drive*. https://www.thedrive.com/opinion/18952/this-is-the-human-driving-manifesto. Accessed on 5 Sept 2021.

Royakkers, L., Est, V., & Rinie. (2015). *Just ordinary robots: Automation from love to war.* CRC Books.

Scheffler, S. (2011). *Death and the afterlife.* Oxford University Press.

Scheffler, S. (2018). *Why worry about future generations?* Oxford University Press.

Schwitzgebel, E., & Garza, M. (2020). Designing AI with rights, consciousness, self-respect, and freedom. In S. Matthew Liao (Ed.), *Ethics of artificial intelligence* (pp. 459–479). Oxford University.

Scott, D. (2012). Geoengineering and environmental ethics. *Nature Education Knowledge, 3*(10), 10.

Singer, P. (2021). Why vaccination should be compulsory. *Project Syndicate.* https://www.project-syndicate.org/commentary/why-covid-vaccine-should-be-compulsory-by-peter-singer-2021-08. Accessed on 5 Sept 2021.

Teitel, R. G. (2000). *Transitional justice.* Oxford University Press.

Open Access   This chapter is licensed under the terms of the Creative Commons Attribution 4.0 International License (http://creativecommons.org/licenses/by/4.0/), which permits use, sharing, adaptation, distribution and reproduction in any medium or format, as long as you give appropriate credit to the original author(s) and the source, provide a link to the Creative Commons license and indicate if changes were made.

The images or other third party material in this chapter are included in the chapter's Creative Commons license, unless indicated otherwise in a credit line to the material. If material is not included in the chapter's Creative Commons license and your intended use is not permitted by statutory regulation or exceeds the permitted use, you will need to obtain permission directly from the copyright holder.

Chapter 9
Designing in Times of Uncertainty: What Virtue Ethics Can Bring to Engineering Ethics in the Twenty-First Century

Jan Peter Bergen ⓘ and Zoë Robaey ⓘ

9.1 Introduction

Among the many lessons that the COVID-19 pandemic has forced upon us is the stark reminder that we cannot assume that the world of tomorrow will be like today. Serious changes to our communal lives can happen quickly and with little warning. When that happens, we need to change with it. But, of course, COVID-19 is not alone in driving such developments. Climate change, geopolitical developments, or disruptive technological innovations present serious moral, political, economic, scientific and/or technical challenges. While none of these can be solved by only technological means, technological innovation will undoubtedly be part of confronting such issues.

Unsurprisingly, this translates into responsibilities for those developing these technologies, especially as calls for morally responsible innovation become increasingly pressing (Brundage & Guston, 2019). However, this puts innovators – including engineers – in a difficult position since making design choices implies having sufficient information to make those choices and the accompanying trade-offs. However, in times of rapid and sometimes unpredictable socio-technical change, the necessary knowledge may be lacking, either because the facts of the matter and/or the normative aspects that should guide decision-making are unclear (van de Poel & Robaey, 2017). In such cases of uncertainty, deliberation about what constitutes good engineering design becomes difficult. Furthermore, given the fact that an innovation's consequences (e.g., in cost-benefit analysis and risk assessment) and

J. P. Bergen
University of Twente, Enschede, Netherlands
e-mail: j.p.bergen@utwente.nl

Z. Robaey (✉)
Wageningen University, Wageningen, Netherlands
e-mail: zoe.robaey@wur.nl

© The Author(s) 2022   163
M. J. Dennis et al. (eds.), *Values for a Post-Pandemic Future*, Philosophy
of Engineering and Technology 40, https://doi.org/10.1007/978-3-031-08424-9_9

norm-based prescriptions (e.g., in engineering codes of conduct) often guide engineering decision-making, it should not surprise us that decision-making becomes more complicated when one or both types of inputs for good engineering decisions are insufficient.

To overcome the deficiencies of consequentialist and deontological approaches to engineering ethics, some have proposed that virtue ethics could help (e.g., Schmidt, 2014) and may be particularly helpful when facing uncertainty (Frigo et al., 2021). This chapter explores these suggestions, strengthening the case for virtue ethics in engineering ethics by showing how it can help deal with different types of uncertainty. We first summarise the case for virtues in engineering ethics (Sect. 9.2), present different types of uncertainty, and compile a list of virtues relevant for engineering (Sect. 9.3). Finally, we present four uncertainty scenarios to analyse the impact of different virtues on engineering decisions (Sect. 9.4).

9.2 The Case for Virtues in Engineering Ethics

The ancient tradition of virtue ethics experienced a revival in the twentieth century, which eventually saw it positioned next to consequentialism and deontology as one of the three major streams of modern ethical thought (Baril & Hazlett, 2019). It comes as no surprise, then, that virtue ethics has since been proposed as an alternative to consequentialist and deontological approaches for various human activities, not least of which engineering. In this section, we summarise the case made for virtue ethics in engineering, focussing on its alleged advantages vis-à-vis its theoretical rivals.

Generally, virtue ethics helps identify engineering as a normative and purposive practice, in turn facilitating an ethical understanding of its activities. Bowen (2009, 2014) proposes we understand engineering as a practice in the sense presented by MacIntyre, i.e., as a "coherent and complex form of socially established cooperative human activity through which goods internal to that form of activity are realised in the course of trying to achieve those standards of excellence which are appropriate to, and partially derivative of, that form of activity, with the result that human powers to achieve excellence and human conceptions of the ends and goods involved are systematically extended" (1981, p. 187). Viewing engineering through this lens, Bowen presents an explicitly virtue-ethical account of engineering. Following his analysis of its internal goods (e.g., technical excellence, cost-effectiveness, safety and especially, the satisfaction of contributing to the flourishing of others) and its external goods (e.g., prestige, wages, economic benefits for others, but most importantly, technological artefacts), he proposes the following end or goal of engineering: "the promotion of the flourishing of persons in communities through contribution to material well-being" (p. 20). To achieve this normative end by attaining engineering's internal goods, engineers need to act ethically. However, the question remains: how is virtue ethics better equipped to help engineers act ethically than its consequentialist and deontological counterparts?

The first such advantage of virtue ethics is increased motivation to practise engineering responsibly. The dominant consequentialist and deontological foundations of engineering ethics aim primarily to prevent misconduct, risks, and disasters. This results in a 'preventive ethics' of engineering (Harris, 2008). However, while preventive ethics may be partially effective in achieving its goals, it lacks "an internal, motivational, and often idealistic element present in professional life that cannot adequately be accounted for by rules" (p. 155). It is this element that is said to be better accounted for and mobilised by virtue ethics, resulting in an 'aspirational' ethics that has a positive rather than a preventive orientation (Bowen, 2009; Harris, 2008; Schmidt, 2014; Steen et al., 2021). This positive orientation also captures some of the zeitgeist we experienced at the beginning of the COVID-19 pandemic. In her essay 'The Pandemic is a Portal' writer Arundhaty Roy (2020) invited readers to think about what kind of world we should aspire to after this historically transformative experience, an invitation that virtue ethics would extend to engineers as well.

On the one hand, this is because virtue ethics may be comparatively better suited to keep engineers pointed towards the normative goal of engineering. This is in part due to the 'modern' origins of consequentialist and deontological ethics since, like engineering, they function according to the paradigm of 'technical rationality' (Schmidt, 2014), which assumes that the knowledge and skill necessary for a practice can be captured in specific and generalisable rules, prescriptions or instructions. This abstraction of ethics from the actual practice and from the persons involved risks aggravating the disconnect between engineers and the goal of their practice (i.e., the flourishing of people), with them considering values like efficiency and technical ingenuity as ends in themselves instead (Bowen, 2009; Schmidt, 2014). Due to virtue ethics' focus on the person's virtues and the connection of professional virtue to the goal of engineering as a practice, it helps minimise this disconnect. On the other hand, virtue could be intrinsically rewarding to virtuous engineers. Indeed, virtue ethics promises to both "identif[y] good behavior and provid[e] the psychological motivation for conforming to that behavior. It may be a more effective prod to achieve good in the world than are less personal calls to maximize utility or conforming to rights and duties, precisely because it is so personal" (Crawford-Brown, 1997 p. 483). This has to do with the nature of virtues as moral dispositions achieved through education and experience. As Aristotle already noted in the Nicomachean Ethics (NE, II.3), both virtue and vice are concerned with pleasure and pain. However, it takes proper education to learn to enjoy the things one ought, to feel delight and pain rightly. It is a sign that a person has achieved virtue that they derive pleasure from exercising it and are characteristically disposed to doing so. This is no less true for professionals than it is for others, even if the virtues demanded by the profession may be specific to it. Indeed, "adhering to professional virtues brings professional satisfaction, just as adhering to personal virtues brings satisfaction to one's personal life" (Harris, 2008 p. 158). The good engineer receives satisfaction from the virtuous exercise of their profession.

The second critical advantage of virtue ethics is that it helps engineers to perform their profession responsibly in a way not covered by its more act-oriented alternatives: "fulfilling an engineer's responsibilities to protect public safety, health, and welfare calls as much for settled dispositions, or virtues, as it does for performing this or that specific action." (Pritchard, 2001 p. 391). This is because those stable dispositions, both personal and professional, are an integral part of professional competence. This makes sense when understanding engineering as a normative practice since virtues are integral to practices that allow practitioners to achieve the practice's internal goods and end (Harris, 2008). One way virtues help the engineer practise their profession responsibly involves a peculiar aspect of virtue as a stable and deeply entrenched trait of character. Following Hursthouse, Harris sees virtue as 'multitrack', involving not simply reason, but "emotions and emotional reactions, choices, values, desires, perceptions, attitudes, interests, expectations and sensibilities." (Hursthouse, 2006 as cited in Harris, 2008 p.156). To cultivate engineering virtues is thus also to attune one's perception, values and sensibilities to those salient aspects of the context of engineering that allow the attainment of engineering's internal goods and goal. As a result, virtuous engineers are more likely to care about and be more sensitive to important aspects of situations and thus more prone to discover them. For example, the rapid development of contact-tracing apps during the COVID-19 pandemic raised difficult trade-offs between privacy, security, public health and a host of other values that led to varied designs. In the Dutch context, for example, a strong emphasis was given to privacy (Verbeek et al., 2020) after consultation with experts. An engineer that is appropriately sensitive to and appreciative of the moral gravity of these trade-offs is also more likely to engineer ethically acceptable applications in these trying times.

Related to these sensibilities comes the idea that virtues, as stable traits of good character, allow the engineer to exercise their powers of discretion and judgement (Harris, 2008), improving decision-making (Sand, 2018), e.g., in selecting the appropriate heuristics in engineering decisions (Schmidt, 2014). Thus, not only does virtue ethics leave more room for the engineer to judge, but being virtuous makes them more capable of doing so. Of course, such virtue has to be cultivated through ample education and experience. However, this focus on education should not be taken as a downside of virtue ethics but rather a call for more disposition-oriented engineering ethics education.

At this point, it is imperative to note that the exercise of virtue both in the engineer's personal and professional life requires it to be done deliberately, with practical wisdom. Multiple authors have emphasised the importance of practical wisdom or *phronesis* for the virtuous engineer and the capability for proper engineering judgement (e.g., Frigo et al., 2021; Harris, 2008; Schmidt 2014; Steen et al., 2021). Aristotle defined practical judgement as "a reasoned and true state of capacity- to act concerning human goods" (NE VI.5). For engineers, this complex virtue would allow them to know what aspects of a situation are relevant, what virtues are appropriate in a given situation, how they fit together, and how they should be exemplified in practice (Athanassoulis & Ross, 2010; Steen et al., 2021). As such, the virtue of practical wisdom may be of particular use when the consequences of an action are not entirely known, or the appropriate norms are not given, i.e., it may help engineers when significant aspects of the case are *uncertain* (Frigo et al., 2021).

In light of the above, the case for virtue ethics in engineering -possibly supplementary to consequentialist and deontological approaches- is extensive and *prima facie* convincing. As such, our goal is not to make this case anew. Rather, we aim to *strengthen* the case for virtue in engineering ethics by further highlighting its compatibility with current changes in engineering – specifically the increased appreciation of uncertainty in engineering work, which might be exactly the type of attitude needed for engineering during and after the pandemic.

9.3   Virtue, Uncertainty and Engineering

The latter of the abovementioned advantages, the usefulness of practical wisdom or *phronesis,* is fitting for a practice like engineering that is both a) inescapably normative and b) supposed to provide practical and acceptable solutions in ever-changing circumstances. These characteristics parallel Aristotle's characterisation of practical wisdom as not only concerned with human goods (NE VI.5) but also not "concerned with universals only – it must also recognise the particulars; for it is practical, and practise is concerned with particulars" (VI.7). This positions practical wisdom as a virtue of careful deliberation about the good that is paramount in exercising virtue more generally (VI.13). Nevertheless, to deliberate well about proper action in a world of particulars is not straightforward because practical wisdom will run up against practical and epistemic limits, i.e., deliberation about practical action is often made more difficult by uncertainty, something the COVID-19 pandemic has proven time and time again. That is, many rapid technological developments were considered while knowledge of their effects or the situation into which they were to be implemented was incomplete, e.g., contact tracing apps, new vaccine technologies, testing kits, and new anti-viral drugs. The complex and ongoing deliberations that these prompted are emblematic of the fact that as the consequences of our technological interventions (and/or our appreciation of them) have come to extend farther in time, space and our lives, uncertainties have likewise grown despite significant efforts to map them better. Unsurprisingly, then, virtue ethicists have explicitly argued for the need to account for uncertainty in virtue ethics as well as the need for virtue ethics in light of uncertainties brought about by rapid (socio-)technological change. In "Seven Traits for the Future" (MacIntyre, 1979), MacIntyre grapples with the question of what traits would be desirable to promote in our society going forward. Interestingly, the virtue that tops his list is the "Ability to Live with Uncertainty". In line with the above, MacIntyre points out that "there are necessary limits to our predictions about the future of technology [...] The answer is clear: we will have to design people with all those traits [...] necessary for living in an unpredictable environment, people with an ability to live with a large lack of certainty about their future" (p. 5). A more developed account of the need for virtues in light of the uncertainty in an increasingly technological world is provided by Shannon Vallor (2016), who makes the case that the unpredictability of technological development and its consequences brings about a state of 'acute technosocial opacity', which in turn requires that we develop the proper 'technomoral virtues' if we are to deal with that opacity well. From her list of twelve technomoral virtues, some stand out as specifically conducive to dealing with the uncertainty involved in our technological future, i.e., those of *humility, courage, flexibility, perspective* and, of course, *technomoral wisdom* (the latter being structurally similar to *phronesis* in more general virtue ethics). The central conclusion from these accounts is clear: deliberating about proper action when we are uncertain about future consequences and normative changes requires appropriately virtuous and wise deliberators. The extremely polarised and uncharitable nature of many societal discussions about COVID-related technological interventions would also seem to indicate a need for more of such virtuous deliberators.

Suppose this is true for society at large. In that case, it should be at least as important for those who have the power to steer the very technological developments that bring about a significant share of the uncertainties we are discussing. As such, it most certainly applies to *engineers*. Some of those who have argued for a role for virtue ethics in engineering (discussed in the previous section) have also made the link with uncertainty explicit. Sand (2018), while discussing the virtues of innovators more broadly, recognises that even responsible innovation processes have significant, unpredictable effects on society (dealing in so-called 'wicked problems). Countering the critique that a virtue-ethical approach would be unable to provide an answer to the question of what innovators should do in such a context, he points out that a) that is equally a problem for deontological and consequentialist approaches in those circumstances, and b) virtues could nevertheless provide "guidance and orientation for becoming a more creditable person and avoiding making moral mistakes. Certain dispositions and capacities help to assess risks properly and, thereby, enhance good decision making" (p. 84). Sand's focus on risk is not coincidental. Innovation generally, and engineering projects specifically, are characterised by the impossibility "to predict absolutely accurately what their consequences will be" (Ross & Athanassoulis, 2010 p. 148). This leads to situations of *risk*, which "involve, of necessity, uncertainty; therefore, the outcomes of one's actions will be uncertain" (Ross & Athanassoulis, 2012 p. 838). This central feature of engineering practice puts a "peculiar ethical burden" on engineers: "the assessment, management, and communication of risk—the very real possibility that engineered projects and products could detract from the material well-being of some people, rather than enhancing the material well-being of all." (Schmidt, 2014 p. 998). As such, we should expect engineers to be well-equipped to deal with such situations, which, according to Ross and Athanassoulis (2012), is enhanced by having virtuous dispositions, including *phronesis*. As such, virtue ethics' focus on the engineer's moral character and their responsiveness to relevant contextual features can help them deal with uncertainty in the form of risks (Athanassoulis & Ross, 2010). However, as Sand's reference to 'wicked problems' already indicates, technical risks are only one instantiation of uncertainty engineers may be confronted in their work. In what follows, we present different types of uncertainty, including normative ones, that engineers may encounter when designing the very technologies that will shape our future and ideally help others flourish. In so doing, we hope to show that uncertainty in engineering is more multifaceted than a singular focus on risk would disclose.

9.3.1 Broadening the Scope of Engineering Uncertainty

In a technical sense, uncertainty represents the lack of probabilistic knowledge for a given event (Doorn & Hansson, 2011). Engineering's preoccupation with risk rather straightforwardly fits this 'technical' sense of uncertainty. However, scholarship in the philosophy/ethics of technology and engineering has gone further in defining types of uncertainties that engineers may encounter and capture the complexity of what is unknown. In this section, we present four types of uncertainty from recent publications from these fields. In each type, uncertainty represents a situation that may hamper engineering decision-making, i.e. design decisions in which the potential risks of new technologies play a defining role.

A first type of uncertainty identified in the engineering ethics literature is *scenario uncertainty* (van de Poel & Robaey, 2017). This type of uncertainty captures the lack of full knowledge about a situation, where different potential ways forward (scenarios) can be imagined based on the available information. Still, we lack the knowledge to reasonably predict how likely these scenarios are to unfold. We often lack complete knowledge for many innovations, but there are some benchmark or a history of use that provide reasonable expectations of a way forward. In scenario uncertainty, this is not the case. Moreover, it is not only that we do not know which scenarios will be likely to happen but -because the form of our knowledge or the lack thereof has normative consequences- also which ones would be more or less desirable. Thus, this type of uncertainty will have normative implications and relate to *epistemic normative uncertainty* (Taebi et al., 2020).

Another type of uncertainty, also related to epistemic normative uncertainty, is *ignorance*: a situation where there is simply no knowledge of some potential consequences of a technological intervention (van de Poel & Robaey, 2017). Here, we don't even know what scenarios we don't know.

There are two more types of uncertainties we include in our analysis. In these, the normative aspects of engineering applications take centre stage. The third type of uncertainty identified in the literature mentioned above is *indeterminacy*, the situation in which causal chains are uncertain, and different actions of different agents could lead to different outcomes that might be unforeseen (van de Poel & Robaey, 2017). This type of uncertainty also evokes *evolutionary normative uncertainty* where, as technology and moral norms co-evolve (Taebi et al., 2020), it is unclear how to normatively assess a technological innovation because of unpredictable moral change. A fourth and final type of uncertainty is *normative ambiguity* representing a disagreement about values and norms (van de Poel & Robaey, 2017). Normative ambiguity can be further specified as *theoretical normative uncertainty*, where different ethical theories will justify different ways of dealing with a problem

(Taebi et al., 2020), or *conceptual normative uncertainty* where the norms themselves allow for different interpretations or prioritizations (Taebi et al., 2020).

Even from this condensed summary, it is clear that uncertainty has many forms and, importantly, includes *normative* unknowns. In light of this, if virtues are supposedly effective at helping engineers deal with uncertainty, and the exercise of virtue aims at the practical and the good, it stands to reason that virtue may be helpful for both epistemic and normative uncertainties. However, whether and to what extent this is the case remains to be seen. As such, this chapter aims to evaluate the usefulness of engineering virtues when faced with various forms of uncertainty in engineering. Before it can do so, however, we must first consider what engineering virtues might be applicable in the first place.

9.3.2 What Are the Engineering Virtues?

Above, we said that this chapter does not aim to make a case for virtue ethics in engineering anew. Likewise, it does not aim to develop a new list of most important engineering virtues, nor is this necessary given the impressive array of engineering virtues presented by others. Table 9.1 offers an extensive overview of engineering virtues as proposed by those making a case for virtue ethics in engineering; virtues that we take as inspiration in analysing the uncertainty scenarios below.

This table structures the overview of engineering virtues differently from how one would find them in the sources from which they were extracted. The presentation of these virtues, their character, and their connection to other virtues and to engineering varied widely across those sources. Thus, the virtues in Table 9.1 have been brought together under a new structure. However, because of these varied foundations, and because we are not taking a strong position on the nature of virtue, Table 9.1's virtue categories should be read heuristically.

For example, it follows the general distinction between moral and intellectual virtues. However, it also has an 'in-between' category that contains virtues that do not neatly fall in either of those two. For example, whether humility is a moral or an intellectual virtue may depend on whether we have outcomes – or motivations based understanding of virtue (Wilson, 2017) and/or the context in which it is acted upon (Bommarito, 2018). Likewise, the virtue of anticipation has been linked to both moral responsibility as well as intellectual virtue in engineering (Steen et al., 2021; Stone et al., 2020). It has even been argued that some virtues can be hybrid, simultaneously moral and intellectual (e.g., hermeneutic justice, see Fricker, 2007). Answering the question of which of these possibilities fits which virtue in the category is beyond the ambitions of this chapter. However, it indicates that exercising these virtues may exhibit both moral and intellectual excellence.

Next, Table 9.1 distinguishes between 'fundamental human virtues relevant for engineering' and 'specifically engineering virtues'. This is simply to indicate that the former would be expected of virtuous persons generally but are also important for the virtuous engineer, while the latter are specific to the practice of engineering.

Table 9.1 Overview of previously recognised virtues relevant for engineering

Fundamental Human Virtues Relevant for Engineering		
Predominantly moral virtues	**Context-dependent or hybrid virtues**	**Predominantly intellectual virtues**
Justice [2, 6, 7] Civic-mindedness/respect for life, law and the public good [1, 4] Compassion, empathy and care [6, 7] Charity/generosity [2, 7] Magnanimity [7] Beneficence [2] Temperance [2] Willingness for self-sacrifice [4] Appropriate ambition [4] Civility [7] Self-control [7] Empowerment [7]	Courage [2, 4, 7, 8] Honesty [1, 4, 6, 7] Integrity [1, 2, 4] Humility [2, 7] Flexibility [2, 7] Perseverance [2, 4] Fortitude/vigor [2] Tenaciousness [5]	Truthfulness [2, 6] Open-mindedness [2] Originality [2] Thoroughness [2] Carefulness [2]
Practical wisdom (*Phronesis*) [6, 7, 8, 9]		
Specifically Engineering Virtues		
'Social' virtues	**Context-dependent or hybrid virtues**	**'Technical' virtues**
Cooperativeness [4, 6, 7] Techno-social sensitivity [3] Respect for nature [3] Commitment to the public good [3] Dedication to safety [8] Willingness to compromise [4] Inclusion and responsiveness [7] Responsible leadership [1]	Anticipation [7] Perspective [7] Dedication/commitment [7]	Creativity/imaginativeness [4, 5, 7] Openness to criticism/correction [4, 8] Accuracy [1] Rigour [1] Curiosity [7] Solution-orientedness [8] Competence/expertise [4] Objectivity [4, 6] Commitment to quality [4] Sensitivity to risk [3] Sensitivity to 'tight coupling' and 'complex interaction' [3] Ability to communicate clearly and informatively [4] Habit of documenting work thoroughly and clearly [4] Seeing the 'big picture' as well as the details of smaller domains [4] Vocation/calling/passion [7]

Sources: [1] Bowen (2014), [2] Crawford-Brown (1997), [3] Harris (2008), [4] Pritchard (2001), [5] Sand (2018), [6] Schmidt (2014), [7] Steen et al. (2021), [8] Van de Poel and Royakkers (2011), [9] Frigo et al. (2021)

This specificity also prompted a redesignation to 'social' and 'technical' virtues, indicating the community-orientedness of the former and the partially practical rather than solely intellectual nature of the latter. Such a division is to be expected for engineers since their being persons and being engineers is not distinct. It is important for engineers to find "continuity and coherence in both professional and personal life. [They] are human persons always and only sometimes engineers" (Bowen, 2014 p. 25).

Interestingly, it would seem that the specific virtue characterisation of the practice of engineering has generally focused on the 'technical' rather than the 'social' virtues, with the moral virtues for engineers being left less specific to the practice. While this is likely partly due to the cognitive nature of engineering work, its practical and normative orientation could be a reason to explore this imbalance further.

Lastly, the central position of practical wisdom in the table is no coincidence. Not only is it one of the virtues most often cited, but when it is, it is given a central, regulative role in the exercise of the engineering virtues as well as in appropriate, context-sensitive engineering judgement (Frigo et al., 2021; Schmidt, 2014; Steen et al., 2021). This, of course, runs parallel to the role that practical wisdom plays in virtue ethics more generally.

Armed with the different types of uncertainty and a list of engineering virtues, the latter's usefulness in dealing with the former can now be investigated. In the next section, we do so by sketching several scenarios based on real situations and technological innovations in which hypothetical engineering professionals face difficult decisions under different conditions of uncertainty.

9.4   Applying the Virtues to Cases of Uncertainty in Engineering

In this section, we present real events and persons to provide a context for discussion. Our subsequent analysis develops *hypothetical* agent-centred considerations grounded in those real events. This allows us to discuss the role of virtues in decision-making by engineers in situations burdened by different types of uncertainty. Each case exemplifies a particular uncertainty situation. However, few real-life cases are likely to be so ideal-typical as to present only one type of uncertainty. Although that does not make their application in our analysis less salient for the purposes of the chapter, we recognise that they do not tell the whole story. Another observation about the cases we present is that they all capture potentially undesirable situations. We acknowledge that uncertainties need not always be about undesirable situations, and in this analysis, we will see that all cases can present desirable and undesirable uncertainties.

Moreover, the cases presented below are all from the life sciences, where the application of recent scientific findings allow translating them to useful engineering applications. The technologies presented thereby add a layer of uncertainty because of their relation to new knowledge not always resulting from traditional science but also of techno-science (Bensaude-Vincent et al., 2011). This allows examining situations where uncertainty is not only an epistemic endeavour but also a moral one. Finally, the cases presented are engineering applications that result from various forms of bio-engineering that relate directly to the pandemic, like mRNA vaccines, or that could come play a role in mitigating effects of the pandemic, like mechanical wombs, and synthetic milk.

9.4.1 Situations of Scenario Uncertainty and Ignorance

9.4.1.1 The mRNA COVID Vaccines

Recently developed COVID-19 vaccines have raised many questions on their potential side effects. Receiving emergency approval by local regulatory bodies, recommendations on their use for specific age groups changed as they were rolled out. The main discourse concerning administering vaccines under emergency approval has been that benefits outweigh the risks of the disease itself. This consequentialist claim was made from a general public health perspective. However, for vulnerable populations, like pregnant people, children, or people with certain existing conditions, the risks and benefits of the vaccines had not been researched when they were first rolled out.

One event that represents **scenario uncertainty**, in particular, is the June 2021 citizen petition on the assessment of mRNA vaccines, led by Dr. Linda Wastila, professor and Parke-Davis Chair in Geriatric Pharmacotherapy at the University of Maryland. Leading the Coalition Advocating for Adequately Licensed Medicines (CAALM) comprised of scientists, clinicians, and patients advocates, their citizen petition to the Food and Drug Administration (FDA) asked for more caution in the full approval of mRNA COVID vaccines (BMJ Opinion, 8 June 2021b). The petition raises eight points of consideration to the Food and Drug Administration. We highlight three of these here: CAALM asks the FDA to provide evidence that the new vaccines will actually benefit vulnerable groups, to research biodistribution of mRNA vaccines, and to further investigate all severe reactions following vaccination. Following the 23 August 2021, FDA decision to grant full approval to the mRNA COVID-19 vaccine by Pfizer, another member of CAALM, Peter Doshi, senior editor of the BMJ, reiterated the need for the 2-year requirement in phase 3 clinical trials in order to "have the science right" (BMJ Opinion 23 August 2021a).

This citizen petition essentially demands less scenario uncertainty by defining specific areas of concern that need further investigation. Reducing scenario uncertainty, in turn, allows reducing **epistemic normative uncertainty**. To make the right choices about vaccine rollout and considering further measures like mandatory vaccination, we need to know the likelihood of different scenarios.

For certain groups of the population, like pregnant people, this goes even further than a situation of scenario uncertainty but rather of **ignorance**. Pregnant people are excluded from medical clinical trials (Smith et al., 2020), understandably so, given the potentially devastating side effects on their future child. In this sense, they are protected, but in the face of the urgency of the COVID-19 pandemic, pregnant people are also at specifically high risks (Wastnedge et al., 2021). Until governments emitted recommendations on vaccination during pregnancy,[1] pregnant people could, in certain cases, elect to receive vaccines. While not all vaccines are safe to administer during pregnancy, like those containing live viral material, mRNA vaccines have the advantage of not containing live virus material and thus presented the option of being administered to pregnant people. One can only imagine the heightened sense of **epistemic normative uncertainty** for this particular population group, with it possibly impacting future generations.

9.4.1.2   What Virtues to Consider for Scenario Uncertainty and Ignorance?

Normally, in drug development, stage-gate models are used to "fail early and fail often", to ensure that whatever drug is produced at the end has higher safety standards (Hjorth et al., 2017). Now, we enter a hypothetical scenario of a vaccine developer: typically, this would be a scientist working on applying knowledge from biochemistry, pharmacology, molecular life science, immunology and so on to the end of making an effective vaccine. For this discussion, our hypothetical scientist engineers a new vaccine and is acutely aware of the concerns raised in the CAALM citizen petition to the FDA. We could even imagine this scientist being pregnant, thus embodying the two types of uncertainties and experiencing the urgency of developing vaccines to fight the pandemic and protect vulnerable groups.

Considering the list of virtues presented above, here are some we could consider relevant and why we think so. Here, the virtue of *justice* seems particularly relevant when making design choices in terms of access to the new vaccines. Here, as MacIntyre points out, we face the challenge of multiple meanings of justice, where he suggests considering the issue of desert (MacIntyre, 1981, p.249). One interpretation we can offer here is that everyone deserves access to health, so this could mean designing vaccines to benefit all groups of the population. This might come at the cost of other internal engineering goods, like efficiency, or run into other challenges like clinical trials regulations. Typically, engineering solutions do not, at first, focus on justice though they can certainly contribute to it. A *just* engineer would aim to realise the normative goal of engineering; whether it is in her ability and power to do something about it is another question.

Here, general virtues of *integrity, honesty and perseverance* can support the *just* engineer. Aiming to act as a just engineer might prove frustrating, and especially in an urgent and business context where outcomes are needed fast. So having virtues of *integrity, honesty and perseverance* can help her remain *just* in such a high-pressure environment. Finally, virtues of *open-mindedness, originality, and thoroughness* can help realise the goal of a *just* engineer, for they help engineer safe and accessible vaccines for all.

---

[1] France and Ireland recommended vaccination from week 16 of pregnancy, whereas Germany recommended waiting for week 20 in certain Länder, and 13 in others, and the Netherlands advised taking it throughout the entire pregnancy.

Looking at specific engineering virtues, *inclusion* and *responsiveness* are a specification of the virtue of justice, accompanied by virtue of *anticipation*, and the virtues of *sensitivity to 'tight coupling' and 'complex interaction'*. We present these as specifications as it is not clear what the exercise of these engineering virtues would amount to without the virtue of justice. For instance, one could excel at anticipating and yet not mobilise it towards justice or human flourishing.

So far, we've discussed scenario uncertainty and ignorance as the same thing. Does ignorance call for different dispositions than scenario uncertainty? With this specific technological intervention, being a just engineer will likely be paramount in either case. Lacking information, conceptions of justice and moral dispositions become increasingly important. As an analogy, consider Rawl's veil of ignorance. A lack of knowledge prompts an increased need for moral virtues of *compassion, empathy and care* (and, arguably, in the case of technology development, *imaginativeness and creativity*).

In this first analysis of relevant virtues in cases of scenario uncertainty and ignorance, it seems that an engineer apt to deal with these situations is a *just* engineer. The other virtues come to support the exercise of justice as a virtue but could be different, depending on the context and type of scenario uncertainty.

9.4.2 Situations of Indeterminacy

9.4.2.1 The Mechanical Womb

In March 2021, a New York Times article heralded the success of mechanical womb research giving birth to thousands of mice embryos (Kolata, 2021). While this article underlines the scientific advantages of studying the development of mice through a mechanical womb, e.g., by pausing development, it also points to potential future applications to human embryos. Needless to say, one can imagine many applications of a mechanical womb for people who struggle with infertility (Berglund, 2021), new avenues to replace problematic issues of surrogacy (Abecassis, 2016), and for increasing chances of survival of preterm babies (Werner and Mercurio, 2021). At the moment, laws prohibit any research on embryos older than 14 days.

In the New York Times article, one of the scientists interviewed on the matter, Dr. Tesar, not involved in the development of the mechanical womb, is quoted saying: "[e]ven assuming they could [grow human embryos], whether that is appropriate is a question for ethicists, regulators and society." In the enthusiasm of this technoscientific achievement, it seems that scientists interviewed in this New York Times piece defer moral judgement and potential future use to other agents in society.

Questions of **indeterminacy** arise at several levels. In this case, we can expect an entanglement of causal events, different decisions of different stakeholders, and changing norms and values concerning this. Here, we list just a few of the indeterminate issues in relation to the mechanical womb (cf. Verbeek, 2008):

- How would society decide to use this technology: to support pre-term babies in neonatal intensive care units or to implant an embryo until birth to remedy infertility and replace surrogacy?
- How would parents and doctors decide when to use the mechanical womb in either case? Who would set the guidelines and based on what norms and values?
- How will parents experience relating to their babies born from a mechanical womb? Would this threaten the integrity of the baby's future? Past and recent controversial interventions with babies have brought international media attention to the cases of Louise Brown, the first baby born from in vitro fertilisation, or Lulu and Nana, the first gene-edited human babies. While today, in-vitro fertilisation is a common procedure supporting many families in their reproductive journey, gene-editing is forbidden, while it could also further support families in bearing viable children in some cases.
- What will become of entire bodies of professionals such as midwives, or doulas, were mechanical wombs to become the norm in reproductive health?

These questions capture potential **evolutionary normative uncertainty.** Of course, we might change our moral views on many of these issues, but we just don't know what kind of possibilities the mechanical womb will afford, even within a pandemic situation like ours. For example, pregnant people with COVID-19 are more at risk of pre-term births than their healthy counterparts (Villar et al., 2021). Recent news reports in the Netherlands indicated an alarming number of unvaccinated pregnant people needing emergency C-sections as early as 24 weeks of gestation (NOS, 2021). Despite the fact that the mechanical womb is still under development, one can readily imagine a use case for it in similar circumstances, with all the normative uncertainties that brings.

9.4.2.2   What Virtues to Consider for Indeterminacy?

Let us now imagine our bioengineer, with a background in developmental biology, or training in obstetrics, designing a carefully balanced environment meant to support the development of a human being.

Bioengineers' choices impact the integrity of the life of children, but also on the relationships to their parents, and potentially on the future of labour for pregnancy care. Decisions on the use of technology will impact technical choices and vice versa. It will require understanding the impact that technical limitations might have on the fundamental questions listed above. This is typically beyond the purview of a designer, as the focus is on optimising technology to help grow healthy babies.

In order to capture the range of effects on different agents, an important set of virtues here are *compassion, empathy and care*. Indeed, designing the mechanical womb is not just a matter of optimisation, but rather a matter of what should this technology afford for our human identities and integrity. With this comes a requirement for intellectual *carefulness*, in order to verify assumptions on these various relationships and inquire on how such a development changes things for a complex set of people, also in the future. Some engineering virtues can further specify how to be careful: for instance, *cooperativeness* with healthcare providers, patient organisations, surrogacy advocates, and various types of prospective parents. Furthermore, in order to be able to be cooperative, virtues of *inclusion and responsiveness*, as well as the virtues of *perspective,* will help give depth to cooperativeness by inviting speaking to a broad range of stakeholders and engaging with relevant moral issues at stake. Another possible engineering virtue that would help support this endeavour is *seeing the 'big picture' as well as the details of smaller domains.*

Therefore, a compassionate, empathetic, and caring engineer could make design choices that accompany indeterminate situations, where norms are bound to evolve.

9.4.3 Situations of Normative Ambiguity

9.4.3.1 Synthetic Maternal Milk

There is certainly a lot of debate on the 'right' way to nourish a newborn with global health recommendations focussing on maternal milk, and in the cases where breastfeeding is not an option due to health or socio-economic factors, formula from cow milk is presented as the next best alternative. A new development in synthetic maternal milk might significantly change these discussions, creating new opportunities for personalized nutrition. Just like the mechanical womb, the case of synthetic maternal milk could also serve as an helpful alternative for mothers infected with COVID-19 to provide superior nutrition to their children while lowering chances of infection.

An in-depth portrait of Dr. Leila Strickland (Kleeman, 2020), founder of Biomilq, a synthetically produced maternal milk, questions how such a development in precision fermentation could disrupt what we understand as a good way to feed the next generations of newborns and even further prevent breastfeeding in public which is a taboo in many western countries (Hauck et al., 2021). If 'breast is best', how will synthetic breast milk change how people perceive the need for breastmilk, or what kind of added price tag will this put on parents who already pay a premium for formula milk derived from cow milk?

Here we can see various layers of **normative ambiguity** that can be captured at an individual but also societal level. From a public health perspective, an innovation like Biomilq presents several advantages: it is more environmentally friendly as it reduces reliance on dairy farming and reflects parents' preferences in diets as it would be customizable. In the beginning, it would probably be a luxury product using a biopsy from the feeding parent. Still, its production could become available at an attractive price point for consumers in the long run. This is a consequentialist perspective on the matter, often preferred in public health recommendations (Markmann et al., 2015). However, synthetic maternal milk could steer parents away from either formula, or breastfeeding. This can hide a host of problems concerning health and justice. For instance, the designers of Biomilq recognise that synthetic milk will never be equivalent to breast milk due to the adaptive nature of breastmilk: becoming more diluted on warmer days, containing antibodies when a child is sick. Another issue is justice. If synthetic maternal milk is superior to formula, how will parents without means have access to this better option? These two issues point to a rights-based approach to health, where the right to health of a child and the right to access healthy food might be put at risk or might create new demands on the health system. Which of these theories will help us deal with these developments? It is hard to say and rather likely that these different reasoning will create conflicts on the issue of feeding children. This is what *theoretical normative uncertainty* captures. Within this, there will likely be *conceptual normative uncertainty* as to how parents and doctors, hospitals, or even international organisations like the World Health Organization see what as healthy or just.

9.4.3.2 What Virtues to Consider for Normative Ambiguity?

Let us now consider the role of a bioengineer, active in precision fermentation and developing synthetic maternal milk. Looking at the personal story of Dr. Strickland and the account of experiencing emotional stress that comes from not being able to breastfeed, an act linked to being a good parent, providing the best nutrition for a child's health, is a laudable motivation. At the same time, the social disruption potential of such an innovation can change many norms of what is desirable for a child's health, from a public and individual perspective, and what is acceptable in public places in terms of breastfeeding.

In order to be able to think about others, whom a disruptive innovation rather do disservice, our bioengineer might have to exercise *beneficence next to justice*. For example, is it a good invention if it's only good for a portion of parents? This could be further specified as cultivating *appropriate ambition* for an invention and exercising the virtue of *humility* in order to evaluate design choices beyond their own goal. This also calls for the intellectual virtue of *carefulness*. Here this could be understood as being careful to make synthetic milk that is good for children but also being careful about wider issues. A common idea of potentially disruptive innovation is 'break things and move fast' where exactly in these cases, carefulness would be advised.

These fundamental virtues can be further specified in terms of engineering virtues. For instance, to think of beneficence and appropriate ambitions, engineers can develop *techno-social sensitivity* and practise *inclusion and responsiveness* to recalibrate these ambitions and notions of the good for others. This also requires the exercise of *anticipation* of effects or *seeing the big picture as well as the details of smaller domains*. Our engineer needs to be open to criticism and correction for this recalibration of ambitions and notions of the good.

Therefore, a beneficent engineer would be equipped to deal with theoretical and conceptual normative ambiguity by virtue of thinking of the good beyond her own experience.

9.4.4 The Regulative Role of Phronesis in Situations of Uncertainty

The cases described above highlight the interaction of different virtues that support the ultimate exercise of fundamental moral virtues. Depending on how authors classify them, these various types of virtues are human virtues, more specific to engineering, moral/social, intellectual/technical, or somewhere in between. This is the main finding from analysing these cases: types of uncertainties do not necessarily require more knowledge of the truth, but more ways of relating to various issues at stake: for instance, the health of unborn children, parents who struggle with infertility, and parents who want to feed their children with the best possible options. Our interpretative analysis did not make the role of practical wisdom explicit, so we would like to return to it here. In order to navigate between different virtues and be able to know which ones could help support the exercise of other virtues: this demands practical wisdom to be regulative of the various types of virtues (see summary Table 9.2).

In the cases presented, we suggest that some of these virtues are the most relevant virtues and that these are supported by other fundamental human virtues that

Table 9.2 Summary of relevant virtues for life science engineering in different situations of uncertainty

Type of uncertainty	Scenario uncertainty: mRNA vaccines	Ignorance: mRNA vaccines during pregnancy	Indeterminacy: the Mechanical Womb	Normative ambiguity: Synthetic Maternal Milk
Relevant virtue	Justice	Justice	Compassion, empathy and care	Beneficence
Phronesis				
Supportive virtues	Integrity Honesty Perseverance Open-mindedness Originality Thoroughness	Compassion, empathy and care	Carefulness	Appropriate ambition Humility Carefulness
Virtue specification	Inclusion and responsiveness Anticipation Sensitivity to tight coupling and complex interaction.	Creativity/ imaginativeness	Cooperativeness Perspective Techno-social sensitivity Inclusion and responsiveness Seeing the 'big picture' as well as the details of smaller domains	Techno-social sensitivity Perspective Inclusion and responsiveness Anticipation, Openness to criticism/ correction Seeing the 'big picture' as well as the details of smaller domains

can then be further specified into engineering virtues. This is consistent with Aristotle's account where moral virtues and practical wisdom depend on each other in order to achieve human goods (NE VI.13). This analysis from relevant virtues to virtue specification is evaluative and interpretative and might not yield the same conclusions in every case of either type of uncertainty. Rather it underlines the necessity of practical wisdom in recognising which virtues to practise when and which virtues to acquire.

Therefore, the deliberate exercise of practical wisdom is primordial to making design choices, as these choices will require the exercise of various virtues, as our cases illustrate. It is not sufficient to be inclusive if the exercise of justice does not accompany this or to be open to criticism if not for the exercise of beneficence. This confirms the important role of practical wisdom in situations of uncertainty (Frigo et al., 2021).

9.5   Conclusion: Virtues for Designing under Uncertainty

To conclude this chapter, we find that moral virtues are paramount to making good design decisions in situations of uncertainty, like many presented by the COVID-19 pandemic. For the cases we evaluate, we conclude that: situations of scenario uncertainty and ignorance call for a just engineer. Situations of indeterminacy call for a compassionate, empathetic and caring engineer. Finally, situations of normative ambiguity call for a beneficent engineer. Because of the nature of the situations described, we found that the most relevant virtues were, in fact, moral virtues. In addition, we find that these will require the exercise of many other virtues of different kinds. Our analysis also indicates that without a central moral virtue and without practical wisdom, there is perhaps no guarantee of goodness in the exercise of other, more technical virtues.

Our analysis reinforces existing scholarship on virtue ethics and engineering and highlights the need for virtues-based approaches therein, with a specific focus on engineering ethics education. New approaches are needed to *complement* the idea that uncertainties can either be avoided or eliminated through more knowledge. Engineering education could also embrace different situations of uncertainties in order to give space for the cultivation of other virtues; or, as MacIntyre writes, we should have the ability to live with uncertainty.

There are further avenues for research that we have not touched upon in this chapter. First, we have not discussed vices corresponding to the virtues we present. Doing so could bring to light virtue conflicts and present a more complete account of designing under uncertainty. Second, we have not explored poietic virtues (Poznic & Fisher, 2021) as it would require a more sustained interaction with empirical material to add to this discussion than this chapter would allow. Nevertheless, doing so would present additional opportunities to deepen philosophers' and ethicists' empirical engagement with design situations under uncertainty. Third, empirical research could broaden the array of design cases and possibly relevant virtues for engineers, which could, in turn, broaden and/or confirm our idea that some virtues lend themselves better to specific types of uncertainties.

We would like to end this chapter by once more stressing the importance of experience and education for shaping responsible engineers. That is, if we want virtuous engineers ready to tackle an uncertain future, we need to think about how to cultivate the necessary virtues through engineering education in pandemic times and beyond.

Acknowledgements This work is part of the research programme "Virtues for Innovation in Practice (VIPs): A Virtue Ethics Account of Responsibility for Biotechnology" with project number VI. Veni.191F.010, financed by the Dutch Research Council (NWO). ZR is also a research fellow in the Ethics of Socially Disruptive Technologies programme.

References

Abecassis, M. (2016). Artificial wombs: The third era of human reproduction and the likely impact on French and U.S. law. *Hastings Women's Law Journal, 27*(1), 3–28.

Aristotle. (2009). *The Nichomachean ethics (Oxford World's classic).* Oxford University Press.

Arundhaty Roy, S. (2020). Arundhati Roy: 'The pandemic is a portal'. *Financial Times.* Retrieved 3 Dec 2021, from https://www.ft.com/content/10d8f5e8-74eb-11ea-95fe-fcd274e920ca

Athanassoulis, N., & Ross, A. (2010). A virtue ethical account of making decisions about risk. *Journal of Risk Research, 13*(2), 217–230. https://doi.org/10.1080/13669870903126309

Baril, A., & Hazlett, A. (2019). The revival of virtue ethics. In K. Becker & I. Thomson (Eds.), *The Cambridge history of philosophy, 1945–2015* (pp. 223–236). Cambridge University Press.

Bernadette, Bensaude-Vincent Sacha, Loeve Alfred, Nordmann Astrid, Schwarz (2011) Matters of Interest: The Objects of Research in Science and Technoscience. *Journal for General Philosophy of Science 42*(2) 365–383. https://doi.org/10.1007/s10838-011-9172-y

BMJ Opinion. (2021a). *Does the FDA think these data justify the first full approval of a COVID-19 vaccine?*, 23 August 2021. https://blogs.bmj.com/bmj/2021/08/23/does-the-fda-think-these-data-justify-the-first-full-approval-of-a-COVID-19-vaccine/.

BMJ Opinion. (2021b). *Why we petitioned the FDA to refrain from fully approving any COVID-19 vaccine this year.* 8 June 2021. https://blogs.bmj.com/bmj/2021/06/08/why-we-petitioned-the-fda-to-refrain-from-fully-approving-any-COVID-19-vaccine-this-year/

Bommarito, N. (2018). Modesty and humility. *The Stanford encyclopedia of philosophy.* Edward N. Zalta (Ed.). https://plato.stanford.edu/archives/win2018/entries/modesty-humility/

Bowen, W. R. (2009). *Engineering ethics: Outline of an aspirational approach.* Springer.

Bowen, W. R. (2014). *Engineering ethics: Challenges and opportunities.* Springer International Publishing.

Brundage, M., & Guston, D. H. (2019). Understanding the movement(s) for responsible innovation. In R. von Schomberg & J. Hankins (Eds.), *International handbook on responsible innovation* (pp. 102–121). Edward Elgar Publishing Limited.

Crawford-Brown, D. J. (1997). Virtue ethics as the basis of engineering ethics. *Science and Engineering Ethics, 3*, 481–489. https://doi.org/10.1007/s11948-997-0049-8

Doorn, N., & Hansson, S. O. (2011). Should probabilistic design replace safety factors? *Philosophy and Technology, 24*(2), 151–168. https://doi.org/10.1007/s13347-010-0003-6

Fricker, M. (2007). *Epistemic injustice: Ethics and the power of knowing*. Oxford University Press.

Frigo, G., Marthaler, F., Albers, A., Ott, S., & Hillerbrand, R. (2021). Training responsible engineers. Phronesis and the role of virtues in teaching engineering ethics. *Australasian Journal of Engineering Education, 26*(1), 25–37. https://doi.org/10.1080/22054952.2021.1889086

Harris, C. E. (2008). The good engineer: Giving virtue its due in engineering ethics. *Science and Engineering Ethics, 14*(2), 153. https://doi.org/10.1007/s11948-008-9068-3

Hauck, Y. L., Bradfield, Z., & Kuliukas, L. (2021). Women's experiences with breastfeeding in public: An integrative review. *Women and Birth, 34*(3), e217–e227. https://doi.org/10.1016/j.wombi.2020.04.008

Hursthouse, R. (2006). *Virtue ethics. In Stanford encyclopedia of philosophy*. Meta-physics Research Lab Center for the Study of Languages and Information, Stanford University.

Jennifer, Berglund New Advances in Transplants and Bioengineering Aid in Replacing the Womb. *IEEE Pulse 12*(2) 12–16. https://doi.org/10.1109/MPULS.2021.3066691

Kelly M., Werner Mark R., Mercurio (2022). Ethical considerations in the use of artificial womb/placenta technology. *Seminars in Perinatology 46*(3) 151521-10.1016/j.semperi.2021.151521

Kleeman, J. (2020). "I want to give my child the best": The race to grow human breast milk in a lab. *The Guardian*, 14 November 2020. https://www.theguardian.com/lifeandstyle/2020/nov/14/i-want-to-give-my-child-the-best-the-race-to-grow-human-breast-milk-in-a-lab.

Kolata, G.. (2021). Scientists grow mouse embryos in a mechanical womb. *The New York Times*, 17 March 2021. https://www.nytimes.com/2021/03/17/health/mice-artificial-uterus.html.

MacIntyre, A. (1979). Seven traits for the future. *The Hastings Center Report, 9*(1), 5–7.

MacIntyre, A. (1981). *After virtue*. University of Notre Dame Press.

Marckmann, G. H., Neema, S., Daniel, S., Strech (2015). Putting Public Health Ethics into Practice: A Systematic Framework. *Frontiers in Public Health* 3 10.3389/fpubh.2015.00023

NOS. (2021). *Steeds meer ongevaccineerde zwangere vrouwen belanden op IC*. Retrieved 3 Dec 2021, from https://nos.nl/nieuwsuur/artikel/2407610-steeds-meer-ongevaccineerde-zwangere-vrouwen-belanden-op-ic

Poznic, M., & Fisher, E. (2021). The integrative expert: Moral, epistemic, and poietic virtues in transformation research. *Sustainability, 13*(18), 10416. https://doi.org/10.3390/su131810416

Pritchard, M. S. (2001). Responsible engineering: The importance of character and imagination. *Science and Engineering Ethics, 7*(3), 391–402. https://doi.org/10.1007/s11948-001-0061-3

Ross, A., & Athanassoulis, N. (2010). The social nature of engineering and its implications for risk taking. *Science and Engineering Ethics, 16*(1), 147–168. https://doi.org/10.1007/s11948-009-9125-6

Ross, A., & Athanassoulis, N. (2012). Risk and virtue ethics. In S. Roeser, R. Hillerbrand, P. Sandin, & M. Peterson (Eds.), *Handbook of risk theory: Epistemology, decision theory, ethics, and social implications of risk* (pp. 833–856). Springer Netherlands.

Hjorth R. L., van Hove Fern, Wickson (2017) What can nanosafety learn from drug development? The feasibility of "safety by design". *Nanotoxicology 11*(3) 305–312. https://doi.org/10.1080/17435390.2017.1299891

Sand, M. (2018). The virtues and vices of innovators. *Philosophy of Management, 17*(1), 79–95. https://doi.org/10.1007/978-3-658-22684-8_7

Schmidt, J. A. (2014). Changing the paradigm for engineering ethics. *Science and Engineering Ethics, 20*(4), 985–1010. https://doi.org/10.1007/s11948-013-9491-y

Smith D. D., Jessica L., Pippen Adebayo A., Adesomo Kara M., Rood Mark B., Landon Maged M., Costantine (2020). Exclusion of Pregnant Women from Clinical Trials during the Coronavirus Disease 2019 Pandemic: A Review of International Registries. *American Journal of Perinatology 37*(08) 792–799. https://doi.org/10.1055/s-0040-1712103

Steen, M., Sand, M., & Van de Poel, I. (2021). Virtue ethics for responsible innovation. *Business and Professional Ethics Journal, 40*(2), 243–268. https://doi.org/10.5840/bpej2021319108

Stone, T. W., Marin, L., & van Grunsven, J. B. (2020). Before responsible innovation: Teaching anticipation as a competency for engineers. In J. van der Veen, N. van Hattum-Janssen, H.-M. Järvinen, T. de Laet, & I. ten Dam (Eds.), *Engaging engineering education: Proceedings of the 48th annual SEFI conference* (pp. 1371–1377). SEFI.

Taebi B., Jan H., Kwakkel Céline, Kermisch (2020). Governing climate risks in the face of norma-tive uncertainties. *WIREs Climate Change 11*(5). https://doi.org/10.1002/wcc.666

Vallor S. (2016). Technology and the Virtues: A Philosophical Guide to a Future Worth Wanting. Oxford University Press.

van de Poel, I., & Robaey, Z. (2017). Safe-by-design: From safety to responsibility. *NanoEthics, 11*(3), 297–306. https://doi.org/10.1007/s11569-017-0301-x

van de Poel, I., & Royakkers, L. (2011). *Ethics, technology, and engineering: An introduction.* Wiley.

Verbeek, P.-P. (2008). Obstetric ultrasound and the technological mediation of morality: A postphenomenological analysis. *Human Studies, 31*(1), 11–26. https://doi.org/10.1007/s10746-007-9079-0

Verbeek, P. P. C. C., Brey, P., van Est, R., van Gemert, L., Heldeweg, M., & Moerel, L. (2020). *Ethische analyse van de COVID-19 notificatie-app ter aanvulling op bron en contactonder-zoek GGD.* https://www.tweedekamer.nl/downloads/document?id=af010293-8393-48dd-b7da-73dd3c6f56beandtitle=Ethische%20analyse%20van%20de%20COVID-19%20 notificatie-app%20ter%20aanvulling%20op%20bron%20en%20contactonderzoek%20 GGD.pdf

Villar, J., Ariff, S., Gunier, R. B., et al. (2021). Maternal and neonatal morbidity and mortal-ity among pregnant women with and without COVID-19 infection: The INTERCOVID multinational cohort study. *JAMA Pediatrics, 175*(8), 817–826. https://doi.org/10.1001/jamapediatrics.2021.1050

Wastnedge, E. A. N., Reynolds, R. M., van Boeckel, S. R., Stock, S. J., Denison, F. C., Maybin, J. A., & Critchley, H. O. D. (2021). Pregnancy and COVID-19. *Physiological Reviews, 101*(1), 303–318. https://doi.org/10.1152/physrev.00024.2020

Wilson, A. T. (2017). Avoiding the Conflation of Moral and Intellectual Virtues. *Ethical Theory and Moral Practice, 20*(5), 1037–1050. https://doi.org/10.1007/s10677-017-9843-9

Open Access   This chapter is licensed under the terms of the Creative Commons Attribution 4.0 International License (http://creativecommons.org/licenses/by/4.0/), which permits use, sharing, adaptation, distribution and reproduction in any medium or format, as long as you give appropriate credit to the original author(s) and the source, provide a link to the Creative Commons license and indicate if changes were made.

The images or other third party material in this chapter are included in the chapter's Creative Commons license, unless indicated otherwise in a credit line to the material. If material is not included in the chapter's Creative Commons license and your intended use is not permitted by statutory regulation or exceeds the permitted use, you will need to obtain permission directly from the copyright holder.

Chapter 10
Confronting Ableism in a Post-COVID World: Designing for World-Familiarity Through Acts of Defamiliarization

Janna van Grunsven and Wijnand IJsselsteijn

10.1 Introduction

The COVID-19 pandemic has brought about a dramatic change in how we interact with others in our everyday activities. Two-dimensional screens and online platforms have profoundly mediated how we work, learn, stay in touch with friends and family, and connect with health care providers and therapists. For many, the pervasive digitalization of our social and practical lives has signified a substantial loss, with the pandemic underscoring that in-person interactions play a key if not constitutive role in well-being. At the same time, a significant number of people have experienced the digitalization of our social and practical lives not as detrimental but precisely as conducive to their overall well-being. In particular, many disabled people and disability rights activists have celebrated the increased accessibility to practical and social spaces enabled by the pandemic-induced embracing of online communication platforms and other digital technologies.[1] In the words of Ashley Shew:

---

[1] We will use identity-first as opposed to people-first language in this paper. In doing so, we are following Elizabeth Ladau's (2015) argument that by "intentionally separate[ing] a person from their disability … it … implies that 'disability' or 'disabled' are negative, derogatory words. In other words, disability is something society believes a person should try to dissociate from if they want to be considered a whole person. This makes it seem as though being disabled is something of which you should be ashamed. PFL [people-first language] essentially buys into the stigma it claims to be fighting."

J. van Grunsven (✉)
Delft University of Technology, Delft, Netherlands
e-mail: J.B.vanGrunsven@tudelft.nl

W. IJsselsteijn
Eindhoven University of Technology, Eindhoven, Netherlands
e-mail: w.a.ijsselsteijn@tue.nl

© The Author(s) 2022
M. J. Dennis et al. (eds.), *Values for a Post-Pandemic Future*, Philosophy of Engineering and Technology 40, https://doi.org/10.1007/978-3-031-08424-9_10

> Many accommodations demanded under COVID-19 were implemented within weeks ....
> These are all things that disabled and chronically ill people have wanted for a very long
> time. I hope that when we've flattened the curve and saved as many people as possible, we
> don't return to a world in which disabled people are ignored (especially when COVID-19
> will probably produce more of us). (Shew, 2020a)

Not wanting to return to the 'old normal,' Shew suggests, that a new post-COVID world should retain many of the now widely implemented technology-enabled forms of access that have benefitted so many disabled and chronically ill people.[2] Yet, as Shew readily acknowledges, we must be cautious about the role of digital technologies in a post-COVID world, and the idea that these technologies straightforwardly promote access. For one thing, disability is often co-opted by technology developers in order to illustrate the alleged societal benefits of their products, raising the concern that the genuine access-enabling potential of digital technologies for disabled people can play into a more problematic "screens everywhere" temptation that is "representative of today's dominant approach to technology design" (IJsselsteijn et al., 2020, p. 37). Furthermore, as Shew explicitly warns, although many digital technologies may have made it easier for disabled and chronically ill people to access a range of spaces and resources, *ableist* biases that (de)value some bodies and minds over others are rampant in tech-development. Without combating these biases – biases that "shape how and what we design" (Shew, 2020b) – the return to the 'old normal' that Shew warns against seems all but inevitable.[3]

Our aim in this chapter is to take Shew's call for technology-supported access *and* her warning against technology's ableist tendencies seriously. Starting from the premise that promoting accessibility and resisting ableism in technology development are morally imperative, our paper discusses two distinct conceptions of accessibility, paired with two conceptions of how access thus understood can be promoted through technology development. The first conception builds off the notion of *affordances*, taken from the field of ecological psychology (Gibson, 1979). Using the pandemic as an illustrative case, we show (Sect. 10.3) that an affordance-based notion of access underscores the link between a person's sense of well-being and their habitual sensorimotor embeddedness in a world that they experience as a space of familiarity – a space in which they know their way around and are able to respond effortlessly to the many perceived possibilities for action that the world affords (Van

---

[2] This is not to say that these communication technologies were not available prior to COVID, but rather that everyone is now forced to resort to them. Arguably, this has created a level playing field of sorts, where quite unlike people's access to physical meeting spaces, and the unequal distribution of social and physical affordances embedded therein, interactants meet in virtual spaces that allow them similar affordance to the other. Of course, due to the video-centric nature of digital communication platforms this point only goes so far.

[3] Shew refers not just to technological accommodations but also to a widespread increase in flexibility when it comes to scheduling, deadlines, etc. A more flexible (and critical) approach to productivity and the organization of time, she argues, is something we all benefit from. This signifies another dimension of how we design daily life that could benefit from the experience and knowledge of disabled persons.

Grunsven, 2020). In Sect. 10.4, we will present *Warm Technology* (IJsselsteijn et al., 2020) as a paradigmatic example of a design-approach aimed at designing for world-familiarity – thus supporting accessibility in one sense of the word. The second conception of accessibility comes from the field of Crip Technoscience (Hamraie & Fritsch, 2019) and underscores technology's potential to create access not by promoting world-familiarity but precisely by creating friction and disruption within habitual practices and ways of perceiving the world – particularly when those practices are ableist. Though these two perspectives may appear to be in conflict with one another our goal is to defend the importance of both. Promoting accessibility, we suggest, involves a readiness to oscillate between two normative imperatives: (1) recognizing how human well-being depends on world-familiarity, which, in turn, can be materialized through design and (2) recognizing how world-familiarity can harbor pernicious biases that can be called into question through material gestures of *defamiliarization* (Bell et al., 2005) with Crip Technoscience providing an important framework for such defamiliarization. By presenting these two perspectives as mutually required in efforts to design for accessibility, and, furthermore, by framing the pandemic as an event that has placed us, en masse, in a defamiliarized position capable of attuning us to the normative significance of world-familiarity, we hope to better enable technologists and laypersons alike to reflectively evaluate if and how a technological innovation may (or may not) be access-promoting, such that it can contribute to a more just post-COVID world.

10.2 Why We Must Foreground and Finetune the Notion of Accessibility in HCI

As mentioned in the introduction, our proposal starts from the premise that promoting accessibility (and resisting ableism) is a moral imperative in technology development. Our focus is on digital technologies and the space of human-computer-interaction. We first want to briefly elaborate on this premise, before delving into the specifics of our proposal.

First, one might question the need to foreground and finetune the notion of accessibility in the field of human-computer-interaction. After all, this field has been emphatically concerned, at least prima facie, with developing participatory and inclusive design methods that place the needs of digital technology-users at center stage. As Cynthia Bennett notes, "accessibility was one of the most popular keywords describing publications at the 2019 CHI Conference on Human-Computer Interaction" (2020). At the same time, she adds that "despite increased momentum, perspectives from the people with disabilities accessible designs purportedly benefit are under-represented, and these absences may negatively impact people with disabilities and the field of professional design" (2020). There is a need, then, to meaningfully claim and cash-out the notion of accessibility for the field.

Still, one might ask, why focus specifically on *accessibility*, as opposed to, say, inclusivity? We believe accessibility is the right value-concept to foreground for several reasons. Firstly, the notion of *inclusivity* (and of *being included*) can evoke associations of disabled people *being invited* or *brought into* some pre-existing space by those who are in a position of power to extend such an invitation (typically the non-disabled technology experts within that space). While communal belonging is arguably a moral good, we wager that it matters how this belonging is achieved in a procedural sense. The notion that one needs to be included into a space by others, for instance in virtue of design-choices made by non-disabled 'experts,' may (implicitly or explicitly) diminish the degree of agency one is credited with. IJsselsteijn et al. (2020) worry that "this starting point can lead to an inherent disempowerment, and an implicit lack of respect in data collection practices towards [the relevant stakeholders] and in the resulting designs made for them" (p. 40). This is particularly problematic in the context of designs intended for disabled users, given the pervasive tendency (both by tech-developers and society at large) to view disabled people as somehow less agential than non-disabled people and merely as the passive users of technology (Shew, 2020b). This tendency in fact discredits a long history of disabled people actively modifying ("tinkering with" or "hacking") the artefacts they rely on in navigating the environment so as to gain *access* to a world that is by and large designed for non-disabled people. Before non-disabled people started to consider the importance of 'including' marginalized disabled people into 'their' spaces, disabled people were in fact already actively creating and claiming access to those spaces – while simultaneously critiquing and challenging the ableist value-system contouring those spaces (Hamraie & Fritsch, 2019).

Second, then, the notion of accessibility is significant because it underscores the environment's constitutive role in the experience of disability and the work involved in making the material and digital built environment more hospitable to all. The notion of inclusivity, with its connotation of 'bringing people in' may leave unquestioned the normative status of the world that technology designers and developers are attempting to include people into. As Shew discusses, a paradigmatic example of an innovation reflective of this perspective is the exoskeleton, which is designed with the purpose of providing *some* paralyzed wheelchair users with the required abilities that would enable them to quite literally step into a world organized around walking. The idea that this world is better and more desirable is itself left unquestioned. Shew has coined the term *technoableism* to capture this phenomenon, which "describe[s] a rhetoric of disability that at once talks about empowering disabled people through technologies while at the same time reinforcing ableist tropes about what body-minds are good to have [in this case body-minds that are upright and ambulant] and who counts as worthy. Technoableists usually think they have the good of disabled people in mind. They do not see how their work reinscribes ableist tropes and ideas on disabled bodies and minds" (Shew, 2020b, p. 43). The notion of accessibility, we wager, reminds us of the idea (also central to the social model of disability) that what disabled people very often need – and indeed very often already bring about through their own active hacking and tinkering – is not necessarily a change to their individual body-minds such that they can participate in 'our' world, but precisely a change to the wider social and material environment. Combatting ableism, then, crucially involves promoting changes in the environment that increase accessibility.[4] As we will suggest in a moment, an affordance-based approach offers resources for fleshing out this idea; making perspicuous how, in our thriving as agents, we rely upon an embodied habitual familiarity with the possibilities for action afforded by the material and digital built environments in which we are embedded.

Of course, in a straightforward sense 'inclusivity' and 'accessibility' are simply concepts that allow for a range of interpretations. Indeed, as Hamraie and Fritsch (2019) note, the notion of access can also be developed in assimilatory directions. However, they highlight that "the etymology of the word *access* reveals two frictional meanings: access as "an opportunity enabling contact," as well as "a kind of attack," adding that "Taking access as a kind of attack reveals access-making as a site of political friction and contestation" (Hamraie & Fritsch, 2019, p. 10). Following this idea, our final reason for focusing on accessibility is that it encourages us to consider how digital technologies may be designed so as to promote accessibility in this critical friction-creating sense.

In sum, what we will be arguing in the remainder of this chapter, is that promoting accessibility through technology design involves an oscillation between (1) appreciating and designing for accessibility understood as world-familiarity and (2) turning to mechanisms of defamiliarization to critically reflect on the habits, biases, and assumptions that are always an ineluctable part of one's world-familiarity.

10.3 An Affordance-Based Take on Accessibility: Lessons from the Pandemic

The idea that worlds, in virtue of how they are designed and built, can be more (or less) accessible to some people than to others can be further deepened via the notion of *affordances*. Affordances, a term coined by ecological psychologist J.J. Gibson, captures the idea that living beings perceive their environment in terms of the practical possibilities for action it affords them as embodied embedded beings (Gibson, 1979; see also Van Grunsven, 2015; Rietveld & Kiverstein, 2014; Dreyfus, 2007). For instance, a chair is built for sitting and a living being whose needs, embodied sensorimotor skills, and socio-cultural practices make sitting desirable, possible, and meaningful will typically directly perceive a chair as *affording-to-be-sat-on*. When, as affordance-responsive beings, our embodied sensorimotor skills are attuned to the artefacts that surround us, we typically know our way about in the

---

[4] Crucially, as Shew (2020b) notes, increased accessibility through changes in the wider social and material environment isn't just what disabled people need. Whereas the exoskeleton is only capable of (potentially) benefitting some wheelchair users, public ramps – as opposed to steps – benefit many people with limited mobility, including people with various disabilities, many aging adults, parents with young children, people who are temporarily injured etc.

world without requiring much effort or reflection. To borrow an example from Van Grunsven (2020), when you enter a crowded subway car you typically do not have to thematize the number of passengers around you to take up an appropriate distance from them, nor do you have to pay attention to the shape of the subway pole in order to be able to grab it and maintain your balance. You effortlessly and habitually negotiate these social and practical affordances. If you had to focus on how to shape your hand in order to grip the subway pole appropriately; how to maintain your balance while the train was in motion; how to maintain an appropriate distance from the other passengers on the train, you would have a hard time directing your thematic attention to other, arguably more meaningful, activities (having a conversation with a friend; reading a book; rehearsing an important conversation with your boss that you plan to have later that day, listening to your favorite music or podcast, etc.).

The effortless pre-reflective embeddedness in practical environments described here depends on a close-coupled match between an agent's embodied sensorimotor skills on the one hand and the material environment on the other hand. Accessibility, we propose, can be understood in terms of this match between the affordances available in a given environment and the embodied skills and capacities that enable situated agents to perceive these affordances as familiar features of a world in which they habitually know their way around. Crucially, this means that not everybody enjoys equal access to the world understood as a space of familiar affordances. Stair cases, door-handles, public bathroom, bicycles, cars, tablets, smart phones, key boards, screens – all these artefacts and features of the technological built environment are designed for certain types of embodied minds (often young adult, neurotypical, able-bodied, and digitally literate) who possess certain (sensorimotor) skills, capabilities and preferences. When this is forgotten or ignored in technology development, seemingly innocuous choices made at the level of design effectively legislate who has access to the world as a space of familiarity.

Many of the habitual flow-like interactions with the environment that people (especially able-bodied people) are typically able to take for granted in the course of everyday living have been disrupted as a result of the pandemic; particularly in its early stages when the world seemed to transform overnight from a place of familiarity into a defamiliarized space. As Van Grunsven has noted:

> Many of the most basic features of our practical environment (door-handles, elevators, public transportation, cash, produce, our mail) seem to warrant a new form of engagement. ... The transformation of social affordances—both in the private sphere and the public domain—has been even more dramatic. Strangers on the street largely afford to-be-shunned or avoided. ... many of us no longer dwell in public spaces the way we used to—we no longer casually grab the subway pole and rely on our habitual know-how to take up an appropriate distance to others. Moving about in public space is now more often than not an effortful endeavor (Van Grunsven, 2020).

By defamiliarizing the familiar, dislodging many of us from our habitual taken-for-granted access to a world of familiarity, the COVID-19 pandemic has (in principle) created the conditions for an appreciation of how our sense of agency and over-all well-being depend on a fluent, close-coupled match between our embodied sensorimotor skills on the one hand and our practical socio-technological environment on the other hand:

> If the unreflective activities that tend to support our more labor-intensive thematic forms of world-directedness now warrant thematic directedness themselves, this creates the condition for a specific kind of fatigue stemming from excessive self-monitoring and of reorienting oneself in a world that has lost some of its immediate action-guiding significance. Flow-like engagements are continuously interrupted by attitudes of distrust towards and detachment from the familiar. To put this in terms familiar to psychologists, the loss of world-familiarity brought about by the pandemic can be understood as a distinct source of *ego-depletion* (Van Grunsven, 2020).

A pandemic-enabled awareness of how our functioning and thriving as agents depends on a close-coupled match between our embodied sensorimotor skills on the one hand and the sociomaterial environment on the other hand can be utilized to design for world-familiarity; serving as an experiential resource that non-disabled technology designers can tap into to understand the normative significance of being embedded in a world-familiarity and the strenuousness that might go along with having to navigate an environment whose affordances resist effortless coupling with one's embodied skills and capacities. That said, we acknowledge that while many may have experienced the kind of loss of world-familiarity described here, that this experience of loss was not distributed equally. While, as discussed in the introduction, many disabled people and disability rights activists have welcomed the digitally-enabled forms of access that the pandemic world has embraced en masse, disabled people have still been among those bearing the brunt of the pandemic and the effects of various COVID-prevention measures (c.f., Wright, 2020). That world-familiarity is not an evenly distributed phenomenon makes it all the more important to introduce it as an explicit goal for technology development and design. In the next section we turn to Warm Technology as one example of what a design approach to promoting world-familiarity might look like.

10.4 Warm Technology: Designing to Support Fragile World-Familiarity

An alternative approach to technology design, termed Warm Technology (IJsselsteijn et al., 2020) has recently been formulated in the context of designing for and with people with dementia. Since loss of world familiarity is central to the phenomenology of dementia, the case of designing for dementia using the Warm Technology approach helps underscore the potential as well as the normative significance of using technology to support world-familiarity and, relatedly, well-being. Furthermore, reflection on how world-familiarity for people with dementia can be supported through digital technologies is particularly urgent during these pandemic times. As social distancing measures have had a particularly devastating impact on people with dementia, it is tempting to turn to digital technologies as quick technological fixes for this pressing social problem (Cheung & Peri, 2021). Our worry is that the pandemic could motivate the development and adoption of digital interventions that fail to incorporate a robust reflection on how exactly digital technologies must be designed such that they genuinely support people in finding or retaining meaningful access to social and practices spaces. To make our case we take a closer look at how Warm Technology approaches the design for people with dementia.

Dementia is not a single disease. It is an overall term that refers to a cluster of symptoms affecting memory, thinking, language, motor abilities, and social abilities, which, taken together, are severe enough to reduce a person's ability to understand and deal with the everyday world. Central to the dementia experience is a loss of world familiarity, which comes in different guises. Dementia, as it progresses, is associated with a loss of temporal and spatial awareness, loss of episodic and semantic memory, loss of cognitive planning and control functions (e.g., not being able to coordinate one's behavior, such as cooking a meal, or making an appointment), loss of language abilities, and loss of sensory-motor functions and skilled behaviors. The familiar slowly becomes strange and confusing. One may get lost on well-traveled routes to and from home, or disoriented in familiar places such as a shopping mall or local park. One may forget words, and names of familiar objects, activities or events. Everyday appliances, such as a remote control or mobile phone, become increasingly opaque and inaccessible. Daily chores, rituals and habits become complex and disorganized. People that were once intimately familiar become mixed up with others or altogether hard to recognize. The trusted may become suspect. Attempts at sense-making – to integrate experiences over time and to form a coherent foundation of one's identity and understanding of the present moment – become unanchored from reality, transforming into a gap-riddled and incoherent patchwork of distorted memories, perceptual hallucinations, and confabulation. Eventually, one may lose all sense of understanding, of self-efficacy, of control over one's environment, and one's own body and mind. This is frequently aggravated by a necessary, sometimes forced, move to a care residence, with its dramatic shift in both physical and social contexts – moving from familiar surroundings to deeply unfamiliar ones. All this may result in feelings of alienation, apprehension, confusion, frustration, loneliness, anxiety, or apathy. These processes do not happen overnight – people live with dementia for years while enjoying a relatively good quality of life. Also, there are significant variations depending on the type and stage of dementia, individual differences, and availability of psychosocial and physical support. Even with progression of the disease, many worthwhile experiences are retained to quite advanced stages of dementia, including appreciation of music, of social company, and affective social touch.

In recent years, the health care technology space has witnessed the development of a wide gamut of digital technological interventions aimed at ameliorating some of the challenges caused by dementia. As we flagged earlier, we can expect that this trend will only accelerate as a result of the pandemic's ubiquitous social distancing measures. Some see this this as a welcome "technology evolution in dementia practice," arguing that "health policy makers, service providers and clinicians should take hold of these innovative opportunities and support the technological transformation of dementia practice in the coming years" (Cheung & Peri, 2021). But the proof is in the pudding and will depend on the types of digital interventions pursued and the manner in which these interventions are designed. Typically, technological interventions used in dementia care settings include ambient assisted living, telecare systems, social robots, and internet of things technologies. The *Warm Technology* approach has emerged as a critical reaction to many of these developments. The underlying problem is that these standard 'cold technology' approaches tend to prioritize what is technologically possible instead of what makes sense from the viewpoint of the lived experiences of people with dementia, whose world-familiarity is increasingly fragile yet crucial to their well-being.

When designing Warm Technology for and with people with dementia, the importance of world familiarity is foregrounded in different ways. First, Warm Technology recognizes the diversity of needs, abilities and resources of people living with dementia. With or without dementia, older adults represent a growing and highly diverse group. Old age is not a uniform stage of life for everyone aged over 65, as some developmental models suggest, rather it is a rich, multiform, non-linear, culturally contextualized and deeply personal process. Furthermore, there is growing cultural and ethnic variation amongst seniors in Western countries. Some are tech-savvy or may have had professional careers that involved tech. Many are well-educated, well-traveled, and in relatively good health. Although clearly the dementia experience will play a role in one's personal identity, experience and outlook on life, it does not define a person. As IJsselsteijn et al. (2020) write: "design efforts to support people living with dementia should not focus on the support, substitution or amelioration of functional decline, but on better ways of affirming old age – enabling people to remain open and attached to the world and to other people, and, as Lynne Segal (2013) so beautifully put it, 'staying alive to life itself'" (p. 33). Technologies designed from a deficiency-first instead of person-first perspective tend to translate into interventions such as large red alarm buttons to be worn as a necklace, tracking devices enabling care-takers to monitor the whereabouts of wandering individuals with dementia, or mobility support ('walkers') designed as medical devices. Such interventions, which, promote a medicalized view of the individual tend to be experienced as stigmatizing and alienating. As Don Norman, himself in his mid-80s at the time of this writing, lamented in a critical essay on technology designed for seniors:

> Despite our increasing numbers the world seems to be designed against the elderly. Everyday household goods require knives and pliers to open. Containers with screw tops require more strength than my wife or I can muster. (We solve this by using a plumber's wrench to turn the caps.) Companies insist on printing critical instructions in tiny fonts with very low contrast. Labels cannot be read without flashlights and magnifying lenses. And when companies do design things specifically for the elderly, they tend to be ugly devices that shout out to the world "I'm old and can't function!" We can do better. (Norman, 2019).

Second, and relatedly, when designing for world familiarity, Warm Technology puts the person's lived experience at center stage, connecting to their personal and family history, their cultural background and upbringing, their local context and community, as well as diversity in literacy and skillsets, technological or otherwise. Familiarity with technology may differ substantially, in part because it will depend on the dominant technology of people's formative years (i.e., one's technology generation – Docampo Rama et al., 2001). In terms of technology design, this may imply referencing familiar form factors and interaction metaphors from the formative years of the elderly person. A recent example of this is the StayTuned radio – a communication system designed by Marjolein Wintermans – den Haan (Wintermans et al., 2017). This 'radio' combines the WhatsApp messaging application with a familiar 60s radio exterior, allowing people to scroll through recorded voice messages of their loved ones using a simple turning knob on a familiar radio interface.

Third, Warm Technology acknowledges the importance of rich multimodal sensory experiences when interacting with the world. Instead of populating the environment with hidden sensors and actuators, touch-screens, virtual agents, or robotic devices, Warm Technology means designing for everyday interactions using everyday objects. This preserves the important affordances of objects and their intuitive relation to the dexterous and perceptual skillsets of a person – easy to recognize and to make sense of. In general, it also implies a preference in designing strong-specific, tangible systems over weak-general, virtual ones – typically steering clear from complicated, multi-layered, multi-purpose ('integrated') systems. In short, Warm Technology focuses on the affordances of familiar objects, and thereby adds to the world familiarity of designed technology interfaces.

Thus, Warm Technology is marked by two key constitutive elements. First, it is born from an emancipatory view of living with dementia. It is to de-emphasize disease and deficiency, and instead focus on the unique identity of the person, on the myriad of ways in which the person inhabits their world as a place of familiarity.

The second essential ingredient of Warm Technology, directly following from the first, is to work closely with people with dementia as part of the design process. Many innovations to date have been designed based on the possibilities of technology (a tech-push approach), or based on inputs from people *around* the person with dementia – for example, family members, informal carers or care professionals. Important and valuable as these perspectives are in their own right, they cannot substitute for the first-person perspective of the person with dementia. Research has shown that different needs, wishes, and requirements emerge depending on the perspective of those involved. The active and continued involvement of people with dementia is of key importance to the design of Warm Technology. In this context participatory practices are proposed and are needed (Suijkerbuijk et al., 2019).

At the same time though, we need to acknowledge that here too, a fundamental tension exists, as noted earlier, of "bringing people in" on the designer's terms. That is, people with dementia are invited to take part in the design process, at the initiative of the designer, and within the value system and implicit assumptions of the design team. This means that the timing of inputs, their nature and expressive bandwidth are, at least in part, enabled and constrained within the design process that is determined by the designer. This observation falls within a larger discourse in research methodology literature on the relation between the researcher(s) and the researched. The privileged position of the researcher in relation to research participants has been a recurrent theme, and perceived asymmetry is both an object of ethical as well as methodological concern.[5]

10.5   On the Importance of Instilling Mechanisms of Defamiliarization in Technology Design

In the previous section we saw that IJsselsteijn et al. (2020) argue for genuine participation of people with dementia in the design of warm technologies, such that the technological interventions designed for and with them align with their particular sensorimotor skills and personal histories, thus meaningfully contributing to their precariously maintained world-familiarity. Similarly, Shew stresses the importance of placing the perspectives and needs of disabled people at center stage: "Instead of imagining the desires of disabled people ... why don't technologists simply ask disabled people what kinds of technological applications we want and need?" (Shew, 2020b, p. 47) Yet, as both Shew and IJsselsteijn et al. recognize, 'simply asking' people isn't as simple as it seems. This is because the types of questions asked, the types of answers given and the importance and meaning attributed to those questions and answers are in part motivated and circumscribed by the wider value systems within which we are embedded. Since ableism is one of such value systems "that all of us participate in, including individual disabled people," co-creation initiatives seem important but not sufficient in technology design efforts to combat ableism and to promote accessibility (Shew, 2020b, p. 46). Bell et al. (2005) argue, for instance, that there is a limit to "user-centered design techniques" when it comes to subverting entrenched pernicious value-systems, because of the emphasis placed on the "current needs and desires" of users. Focusing not on ableism but on patriarchy qua value-system, they maintain that:

> Gender assumptions about labor may be built into technology and reinforce stereotypes about who in the home should do what .... Designers have an opportunity to alter these built-in gender assumptions and thereby support different patterns of behavior. This strategy runs counter to user-centered design techniques because it proposes to design not for users' current needs and desires, but to shape alternative needs, desires, and behaviors through design (Bell et al., 2005, p. 168).

The specific strategy focused on shaping "alternative needs, desires, and behaviors" that Bell et al. are referring to is one of *defamiliarization*, which, compels designers "to examine *their automated perceptions of that which is so familiar* that it seems natural and so unquestionable (Bell et al., 2005, p. 151, our italics). By re-contextualizing and reframing "the affordances" of familiar everyday use-objects, such as "door handles, faucets, filing cabinets," we can make them "strange" and

---

[5] This is not to say, however, that the researched do not bring their own agenda to the research situation.

"defamiliarize[e] the familiar" (Bell et al., 2005, p. 153, referring specifically to Donald Norman's *The Psychology of Things*).

Bell et al. focus on literary, textual, ethnographic techniques of defamiliarization capable of offering "a lens to help us see our own design practices in a new light" (p. 154). Thus, they propose that defamiliarization is "available as a strategy to anyone with access to a pen and paper, or more likely, a keyboard and a monitor. Defamiliarization is not tremendously difficult to achieve and most of us have done it before. It is essentially a rich description which renders strange the familiar" (p. 169–70).[6] This can bring into view the pernicious dimensions of our habitual ways of inhabiting our world of familiarity.

Alongside these ethnographic techniques, we have already presented the pandemic as an event that has viscerally exposed most of us to the experience of defamiliarization. To capitalize on this experience as a resource for access-promotion, we furthermore want to highlight the powerful mechanisms for defamiliarization that have been forged by disability activists themselves, contributing to the field of Crip Technoscience. Crip refers to the "anti-assimilationist position that disability is a desirable part of the world, and "technoscience," refers to "the co-production of science, technology, and political life," i.e. the ways in which our scientific and technological endeavors both form and are formed by shared conceptions of the good life and communal membership (Hamraie & Fritsch, 2019, p. 2). As we mentioned in Sect. 10.2, Crip Technoscience proposes to understand "access as friction" or "as a kind of attack" which "reveals access-making as a site of political friction and contestation" (Hamraie & Fritsch, 2019, p. 10). Furthermore, it foregrounds the long history of disabled agents of access-making, where disabled people have actively hacked, altered, tinkered with sociomaterial environments catered towards 'able-bodied' world-familiarity to not only make these environments more conducive to disabled forms of inhabiting the world, but also to explicate and critique entrenched habitual and often ableist ways of experiencing the world, promoting "practices of critique, alteration, and reinvention of our material-discursive world" (Hamraie & Fritsch, 2019, p. 1).

For instance, Collin Kennedy's act of "protesting hospital parking prices by filling the pay-slot on a parking meter with spray foam," defamiliarizes habitual taken-for-granted capitalistic norms of efficiency dictating what constitutes as a normal pace for moving through the world (Hamraie & Fritsch, 2019, p. 12). For another powerful example that illustrates this form of critical access-promotion through Crip Technoscientific acts of defamiliarization, consider "Deep Sea Diving … in a

---

[6] Many qualitative traditions attempt to minimize the distance between researcher and research participant. A particular example, from design research with elderly communities, is the development of the cultural probe method as a way to rebalance this negotiation, and to subvert the roles of the designers and those "to be designed for". Cultural probes are themselves designed to allow for more agency on the part of the participant – more expressive and creative ability, choice and freedom whether, when and in what ways to partake (See Gaver et al., 1999). Here too, defamiliarization techniques could play an important role, in particular to uncover value systems and default implicit assumptions in how to design for and with people living with dementia.

Wheelchair" – a TED talk in which artist and disability rights activist Sue Austin presents the various ways in which she has altered her wheelchair in order to claim her visibility in social space by challenging people's implicit habitual ways of seeing what wheelchairs afford. Seeking for new narratives to reclaim her identity, Austin purposely "transform[s] perceptions by revisiting the familiar." Among other things, Austin turns her wheelchair (or power chair, as she prefers) into a deep sea diving device. As viewers of Austin's work watch her explore the ocean's corals in her under-water power-chair, arms spread wide, she wagers that

> In that moment of them seeing an object they have no frame of reference for or so transcends the frames of reference they have with the wheelchair they have to think in a completely new way. .... For me this means that they are seeing the value of difference, the joy it brings, when instead of focusing on loss or limitation, we see and discover the power and joy of seeing the world from exciting new perspectives. For me the wheelchair becomes a vehicle of transformation. ... Because nobody's seen or heard of an underwater wheelchair before ... creating this spectacle is about creating new ways of seeing, being and knowing. (Austin, 2012)

Though ableism as a pernicious value-system has been materialized into the world through a wide range of technological artefacts and sociotechnical systems, it is also through the tweaking of artefacts and the disruption of sociotechnical systems that entrenched ways of seeing disabled people and perceiving our everyday world of familiarity can be called into question and new unfamiliar ways of imagining the world can open up. We wager that the mechanisms of defamiliarization offered by the field of Crip Technoscience, positioned as forms of access promotion, provide a powerful resource for technology developers and designers who follow the premise of our argument, namely that promoting accessibility (and resisting ableism) is a moral imperative in technology development. Finally, a commitment to Crip Technoscience's mechanisms of defamiliarization can be reinforced by tapping into what we have presented as an important phenomenological feature of the pandemic; namely the sense in which the pandemic can be understood as a mass-scale event of defamiliarization, confronting many of us with the intimate but often taken-for-granted link between well-being and having access to a world of familiarity.

10.6 Conclusion

In his commencement speech 'This is Water", David Foster Wallace offers the following anecdote:

> There are these two young fish swimming along, and they happen to meet an older fish swimming the other way, who nods at them and says, "Morning, boys, how's the water?" And the two young fish swim on for a bit, and then eventually one of them looks over at the other and goes, "What the hell is water?" ... The immediate point of the fish story is that the most obvious, ubiquitous, important realities are often the ones that are the hardest to see and talk about (2005, p. 2).

Wallace calls on us to develop "simple awareness — awareness of what is so real and essential, so hidden in plain sight all around us, that we have to keep reminding ourselves, over and over: "This is water, this is water." It is unimaginably hard to do this ... day in and day out" (2005, p. 8). This difficulty applies to everyone, including those of us who live our lives as technology developers and who, in this capacity, "are the unacknowledged legislators of our technological age" (Winner, 1990, p. 59).

What Wallace calls water, we have called world-familiarity and we have argued that inhabiting the world as a place of familiarity plays an integral role in our thriving as agents – the pandemic, which has pervasively disrupted people's world-familiarity, has underscored as much. Of course, many disabled and chronically ill people have always been aware of this as they confront a world that is, for the most part, neither designed for them nor by them. That world-familiarity is deeply central to well-being is recognized in the Warm Technology approach, which aims to promote accessibility through technology development. However, because world-familiarity turns on the habitual, because, in Wallace's words "it is unimaginably hard" "to keep reminding ourselves ...' This is water," designing for world-familiarity demands acts of defamiliarization, through which we critically examine whose world-familiarity we are in fact designing for to. As such, we have suggested that promoting accessibility involves a readiness to oscillate between two normative imperatives: (1) recognizing how human well-being depends on world-familiarity, which, in turn, can be materialized through design and (2) recognizing how world-familiarity can harbor pernicious biases that can be called into question through material gestures of *defamiliarization* (Bell et al., 2005). While the pandemic itself has offered many, if not all, of us a visceral experience of defamiliarization that can serve as a reminder in endeavors of access-promoting technological interventions, we also need tangible mechanisms and frameworks that can guide such projects. As such, we have presented Crip Technoscience as an important resource for defamiliarization – a resource that doesn't frame disabled people as waiting to be included in 'our' world of familiarity, but that actively disrupts some of 'our world's' basic organizing biases, assumptions and value-commitments.

By presenting these two perspectives as mutually required in efforts to design for accessibility, we hope to better enable technologists and laypersons alike to reflectively evaluate if and how a technological innovation may (or may not) be access-promoting, such that it can contribute to a more just post-COVID world; a world where we can all not merely survive, but thrive as precarious embodied world-dependent beings.

Acknowledgments This work is part of the research programme Ethics of Socially Disruptive Technologies, which is funded by the Gravitation programme of the Dutch Ministry of Education, Culture, and Science and the Netherlands Organization for Scientific Research (NWO grant number 024.004.031).

References

Austin, S. (2012). *Deep sea diving ... in a wheel chair*. TED talk. Accessed on 17th Nov 2021. https://www.ted.com/talks/sue_austin_deep_sea_diving_in_a_wheelchair?language=en

Bell, G., Blythe, M., & Sengers, P. (2005). Making by making strange: Defamiliarization and the design of domestic technologies. *ACM Transactions on Computer-Human Interaction, 12*(2), 149–173. https://doi.org/10.1145/1067860.1067862

Bennett, C. (2020). *Toward centering access in professional design*. Doctoral dissertation. https://digital.lib.washington.edu/researchworks/bitstream/handle/1773/45406/Bennett_washington_0250E_21264.pdf?sequence=1. Accessed on Nov 2017.

Cheung, G., & Peri, K. (2021). Challenges to dementia care during COVID-19: Innovations in remote delivery of group cognitive stimulation therapy. *Aging and Mental Health, 25*(6), 977–979. https://doi.org/10.1080/13607863.2020.1789945

Docampo Rama, M., de Ridder, H., & Bouma, H. (2001). Technology generation and age in using layered user interfaces. *Gerontology, 1*(1), 25–40. https://doi.org/10.4017/gt.2001.01.01.003.00

Dreyfus, H. L. (2007). The return of the myth of the mental. *Inquiry, 50*(4), 352–365. https://doi.org/10.1080/00201740701489245

Gaver, B., Dunne, T., & Pacenti, E. (1999). Design: cultural probes. *Interactions, 6*(1), 21–29. https://doi.org/10.1145/291224.291235

Gibson, J. J. (1979). *The ecological approach to visual perception*. Houghton Mifflin.

Hamraie, A., & Fritsch, K. (2019). Crip technoscience manifesto. *Catalyst: Feminism, Theory, Technoscience, 5*(1), 1–34. https://doi.org/10.28968/cftt.v5i1.29607

IJsselsteijn, W. A., Tummers-Heemels, A., & Brankaert, A. (2020). Warm technology: A novel perspective on design for and with people living with dementia. In R. Brankaert & G. Kenning (Eds.), *HCI and design in the context of dementia, human computer interaction series*. Springer Nature.

Ladau, E. (2015). *Why person-first language doesn't always put the person first*. https://www.think-inclusive.us/post/why-person-first-language-doesnt-always-put-the-person-first. Accessed on 5th Oct 2021.

Norman, D. (2019). *I wrote the book on user-friendly design. What I see today horrifies me. Fast Company*. https://www.fastcompany.com/90338379/i-wrote-the-book-on-user-friendly-design-what-i-see-today-horrifies-me. Accessed on 1st Nov 2021.

Rietveld, E., & Kiverstein, J. (2014). A rich landscape of affordances. *Ecological Psychology, 26*(4), 325–352. https://doi.org/10.1080/10407413.2014.958035

Shew, A. (2020a). Let COVID-19 expand awareness of disability tech. *Nature*. https://www.nature.com/articles/d41586-020-01312-w. Accessed 17th Nov 2021.

Shew, A. (2020b). Ableism, technoableism, and future AI. *IEEE Technology and Society Magazine, 39*(1), 40–85. https://doi.org/10.1109/MTS.2020.2967492

Suijkerbuijk, S., Nap, H. H., Cornelisse, L., IJsselsteijn, W. A., de Kort, Y. A. W., & Minkman, M. M. N. (2019). Active involvement of people with dementia: A systematic review of studies developing supportive technologies. *Journal of Alzheimer's Disease, 69*(4), 1041–1065. https://doi.org/10.3233/JAD-190050

van Grunsven, J. (2015). *Bringing life in view: An enactive approach to moral perception* (Doctoral dissertation, The New School).

Van Grunsven, J. (2020). Perceptual breakdown during a global pandemic: Introducing phenomenological insights for digital mental health purposes. *Ethics and Information Technology*. https://doi.org/10.1007/s10676-020-09554-y

Wallace, D. F. (2005). *This is water: Some thoughts, delivered on a significant occasion, about living a compassionate life*. Commencement speech given at Kenyon College on May 21st 2005.

Winner, L. (1990). Engineering ethics and political imagination. In P. T. Durbin (Ed.), *Broad and narrow interpretations of philosophy of technology* (pp. 53–64). Kluwer Academy Publishers.

Wintermans, M., Brankaert, R., & Lu, Y. (2017). Together we do not forget: co-designing with
people living with dementia towards a design for social inclusion. In *proceedings of the design
management academy 2017* (Vol. 2, pp. 767–782). International conference, Hong Kong.

Wright, R. (2020). Who is "worthy"? Deaf-blind people fear that doctors won't save them from
the coronavirus. *The New Yorker*. https://www.newyorker.com/news/our-columnists/who-is-
worthy-deaf-blind-people-fear-that-doctors-wont-save-them-from-the-coronavirus. Accessed
on 17th Nov 2021.

Open Access   This chapter is licensed under the terms of the Creative Commons Attribution 4.0
International License (http://creativecommons.org/licenses/by/4.0/), which permits use, sharing,
adaptation, distribution and reproduction in any medium or format, as long as you give appropriate
credit to the original author(s) and the source, provide a link to the Creative Commons license and
indicate if changes were made.

The images or other third party material in this chapter are included in the chapter's Creative
Commons license, unless indicated otherwise in a credit line to the material. If material is not
included in the chapter's Creative Commons license and your intended use is not permitted by
statutory regulation or exceeds the permitted use, you will need to obtain permission directly from
the copyright holder.

Chapter 11
Understanding Risks and Moral Emotions in the Context of COVID-19 Policy Making: The Case of the Netherlands

Sabine Roeser

11.1 Introduction

A Chinese colleague said in early March 2020: "My Chinese friends and I are very concerned. We have seen what happened in China. This is not just any virus. We don't understand that Western countries do not take stricter measures." In the beginning of the COVID-19 pandemic, many people in Western countries, including politicians, took pride in publicly stating that they were not worried about the Sars-Cov-2 virus. For example, Dutch prime minister Mark Rutte insisted to shake hands with a journalist when asked about safety measures concerning social distancing. In various respects, the Netherlands were late in introducing measures that were already urgently recommend by the WHO (NOS, 2021d). Likewise, some virologists assured the public that COVID-19 was nothing more than just another flu virus from which they could not get sick because only people with vulnerable health conditions were susceptible. Potential worries were explained away. For example, in February 2020, in pieces for major news outlets, psychologists Paul Slovic and Daniel Kahneman as well as legal scholar Cass Sunstein claimed that supposedly exaggerated reactions to the COVID-19 virus could be elucidated by an opposition between reason and emotions. They argued that people's perception of risk is driven by 'irrational' emotions based on which people close themselves off from scientific facts (Fisher, 2020; Sunstein, 2020). Slovic and Sunstein stated that worries about COVID-19 could be readily explained via this framework. A few weeks later, almost the whole world went into lockdown, but it was already too late to stop this supposedly harmless virus, the pandemic was a fact, and the rest is history.

If these experts and politicians were wrong in their initial assessment of the virus, could they also be wrong about their dismissal of emotional responses that

S. Roeser (✉)
Delft University of Technology, Delft, Netherlands
e-mail: S.Roeser@tudelft.nl

© The Author(s) 2022
M. J. Dennis et al. (eds.), *Values for a Post-Pandemic Future*, Philosophy of Engineering and Technology 40, https://doi.org/10.1007/978-3-031-08424-9_11

did highlight concerns about the virus at an early stage? And could we learn important lessons from this, leading to more appreciation of emotions and underlying ethical values and concerns? This is the idea I will pursue in this chapter.

11.2   COVID-19 and Emotions

In his initial speeches about COVID-19 policy, Dutch Prime Minister Mark Rutte repeatedly emphasized the importance of listening exclusively to medical-scientific experts, particularly the virologists and modellers of the National Institute for Environmental Studies (RIVM) and the Outbreak Management Team (OMT). He explicitly said that we should not listen to historians and lawyers, for example, and that we should not "philosophize". Closing the schools in mid-March 2020 wouldn't have been scientifically necessary, but here people have "voted with their feet" (NOS, 2020). Interestingly, a few weeks later, Rutte mentioned that he had been pondering on "dilemmas". However, he still explicitly stated that he would only listen to medical and virological advice, thereby dismissing possible expertise on addressing dilemmas, such as from philosophers and social scientists.

Notably, many people were (and still are) worried because of the impact that policies may have on public health, the economy, and the way of life people are used to. Some are concerned whether it is responsible to send their children to school or to go outside without a mask. Others are angry because they perceive the measures as too strict and because their income is at stake. Until vaccines were available in early 2021, many older people in care homes languished in loneliness during various lockdowns because their loved ones were not allowed to visit. The question is how policy makers and politicians should deal with such emotions and worries, and how philosophical research may shed light on this. In this chapter, I will argue that emotions such as empathy and compassion, as well as resentment and concern, can help to make critical moral dilemmas explicit and thereby contribute to taking moral considerations into account when policy decisions are made about virus-restriction measures.

Obviously, it is crucial to uncover the relevant scientific facts to make important decisions on dealing with a pandemic. But I will argue in this chapter that addressing the COVID-19 crisis and making decisions about trade-offs between different risks is not just a matter of gathering scientific information and listening to scientists, as crucial as that is. Scientific information is *necessary* to make assessments and policy decisions in such a crisis situation, but not *sufficient*. We also have to take into account societal and ethical considerations, which requires explicit ethical reflection and attending to emotions. This argument is grounded in my philosophical approach, according to which emotions can play an important role in ethical reflection (e.g. Roeser, 2011, 2018).

My ideas go against the dominant scientific and political approaches to risk and emotion. As mentioned above, scholars such as Paul Slovic (2010), Cass Sunstein (2005) and Nobel Prize winner Daniel Kahneman (2011) think that all kinds of misunderstandings about risks and statistics can be explained by a contradiction between reason and emotion, so-called 'Dual Process Theory' (DPT). According to DPT, we process information via two distinct systems, "system 1" versus "system 2". System 1 is said to be based on emotion and intuition, and while fast, it is unreliable. System 2 is based on rationality and analytical thinking. System 2 is slower but much more reliable than system 1. According to Kahneman, Slovic, Sunstein, and other psychologists and decision theorists, all kinds of misunderstandings about risks and statistics can supposedly be explained by this: people respond emotionally in their risk perceptions (system 1) and therefore close themselves off to scientific facts (which require system 2 processes).

However, many emotion researchers from psychology and philosophy reject the reason-emotion dichotomy that underlies Dual Process Theory. Instead, they have developed so-called cognitive theories of emotions. The renowned Dutch psychologist Nico Frijda (Frijda, 1986) considered emotions crucial for our appraisals and actions. Philosophers Robert Solomon (1993), Martha Nussbaum (2001) and Bob Roberts (2003) have argued for the importance of emotions for our moral thinking. The neuropsychologist Antonio Damasio (1994) has shown that people who seize to have emotions due to specific brain damage can no longer make practical and moral decisions. These ideas give us a very different and much richer understanding of emotions: emotions are not by definition at odds with rationality as dualistic views of emotion and rationality, such as DPT, entail. Rather, emotions can be an important source of moral reflection and deliberation (Roeser, 2011; Furtak, 2018). Emotions can point to what morally matters. Of course, emotions can also be misguided, but that holds for all our sources of insight. Instead of dismissing emotions, we should see them as an important source of ethical reflection in the context of risks. Emotions can draw attention to important ethical considerations that are frequently overlooked in quantitative, STEM-based approaches to assessing risks. Emotions such as sympathy, compassion, care, and feelings of responsibility can highlight ethical concerns such as justice, fairness, and autonomy. In my previous work, I have argued that these ideas can shed a different light on the role of emotions in decision making about risks, primarily in the context of technological risks (Roeser, 2018). In the remainder of this chapter, I will apply these insights to decision making about COVID-19.

11.3 COVID-19, Risks, Uncertainty, Complexity, and Ethics

The dominant approaches to decision making about risk view emotions as a source of irrationality. A standard approach to decision making about risks is, therefore, to rely solely on scientific expertise. This is what I would call the "technocratic approach"; quantitative information is guiding, public concerns are dismissed as irrelevant. A common alternative strategy is what I call the "populist approach"; here, the public's will is seen as leading. Even if the public's will is attributed to supposedly irrational emotions, it is still followed, either for seemingly democratic or pragmatic (instrumental) reasons to avoid public opposition (cf. e.g. Loewenstein et al., 2001, De Hollander & Hanemaaijer, 2003). However, both approaches fall short because they do not take emotions and underlying values seriously. In neither approach is there a genuine dialogue and deliberation about the values that are at stake (Roeser, 2018). Technocratic risk approaches rely solely on descriptive information and consequentialist methods such as risk-cost-benefit analysis. But such approaches contain implicit and often problematic value judgments. Only net impacts at a high level of aggregation are considered, and often only a limited type of impact, such as the number of deaths. For example, issues such as justice, fairness and autonomy are usually overlooked in such approaches, as are long-term consequences for health and (psychological) well-being (cf. Roeser, 2006; Asveld & Roeser, 2009). Let us begin by zooming in on the ethical intricacies of decision making under risk and uncertainty. After this, in the following section, I will argue that policy measures to combat COVID-19 are intrinsically value-laden. I will then proceed to explore how emotions can contribute to highlighting these ethical issues.

As mentioned above, at various crucial moments, the Dutch government explicitly stated that they would follow the technocratic approach by directly acting on the advice of the STEM-based RIVM and OMT. It can be argued that it is not the task of such STEM-based policy organizations to include ethical considerations in their recommendations. STEM-based approaches are limited to discerning the facts relating to the transition of COVID-19. However, this means that there is an important, unaccounted for 'is-ought gap'[1] between descriptive STEM data and policy decisions, which also have important normative dimensions. One could solve this by having the following separate steps: first, gather the descriptive information, and then have an intermediate step of ethics evaluation before policymakers make decisions based on both steps. However, things are even more complicated: descriptive research also involves normative assumptions, e.g. concerning how to measure, assess and compare data, specifically in the context of risk and uncertainty (Roeser et al., 2012). To account for these issues, even STEM-based councils should include social science experts (to account for the impact and role of society) as well as ethicists to point out implicit ethical and other normative assumptions, highlight ethical dilemmas, and provide for explication of ethical considerations to make these transparent and object of critical deliberation by policymakers and societal stakeholders. This is the case with various governmental advisory boards in the Netherlands, e.g. at the COGEM (committee on genetic modification), as well as the Dutch Health Council, which also provides advice concerning e.g. COVID vaccinations. Ethicists have also been involved in the development of COVID-19 track-and-trace apps. Despite this, ethicists have not yet been involved in decision making on the policy measures, even though these measures include many pressing ethical considerations. This means that these ethical considerations have either been ignored, not made explicit, or dealt with haphazardly without consulting ethics experts' relevant

---

[1] The 'is-ought gap' refers to the issue that one cannot derive a normative conclusion ('ought') from solely descriptive information ('is') (cf. Hume, 1975 [1739–40], Moore, 1988 [1903], Prichard, 1912 etc. for diverging analyses of the implications of this gap).

expertise. This is not to say that ethicists should have the final word on such vital issues. Still, they could play an important role in explicating ethical considerations and highlighting potentially ethically problematic decisions. I will illustrate this in what follows by discussing various aspects of Dutch COVID-19 policies.

There are methodological issues of risk, uncertainty, and complexity that give rise to ethical issues. The measures policy makers implement directly impact the development of the pandemic. Furthermore, numerous actors are involved, and the virus may develop in unforeseen ways, with scientific knowledge lagging behind. This can lead to complex interaction effects, high uncertainty, and a lack of predictability. As pointed out by the critical 'Red Team', COVID-19 should be seen as a case of complexity, requiring different decision-making approaches than conventional, more predictive types of risk. The Red Team was an interdisciplinary team of Dutch scientists (from STEM as well as the social sciences) that criticized the approach of the Dutch government to deal with the pandemic. In 2020, the Red Team strongly influenced Dutch public opinion via social media, as well as being consulted at certain stages of the pandemic by the government. However, they were silent for most of 2021. They recently announced that they had decided to stop working altogether, as their advice was largely ignored and was fundamentally at odds with the strategy of the Dutch government. The Red Team advised to keep infection rates low via early lockdown measures and tracking and tracing. In direct contrast, the Dutch government has from the beginning followed a strategy that primarily steers at preventing the health care sector from getting overburdened, in the meantime being reluctant to employ safety measures.

In recent reports, it has been argued that the initial Dutch, UK, and Swedish approaches to strive for so-called heard immunity were irresponsible, infeasible from the start, and eventually led to thousands of unnecessary deaths. Interestingly, the Dutch government later denied having had such a strategy. Instead, they claimed that they only wanted to achieve herd immunity as a side effect, not as a goal in itself. However, publicly available information such as press communications from the early days of the pandemic as well as internal documentation clearly shows that this was the initial strategy in the Netherlands (cf. NOS, 2021a, b, c). Such wavering communications obviously do not contribute to public trust, which is already a delicate issue given the controversies about different COVID-measures and the various other scandals that the Dutch government and political institutions are currently involved in.[2]

Ethical decision-making about risks presents us with different challenges than ethical decision-making about options where the outcomes are easily predictable or

---

[2] The Dutch government fell in January 2021 due to the so-called 'toeslagenaffaire', i.e. a more than 15 year long systematic tax scandal based on racist and other biases, with widespread consequences for numerous people, and continuously growing evidence of the failure of the rule of law. Since then, various other scandals have emerged that would presumably have led to the falling of the government, if it hadn't stepped down already. Despite elections in March 2021, as of December 2021, the date of finalizing the writing of this chapter, there is still no new government, and the most likely new government is a continuation of the previous coalition.

even fixed, as argued by the Swedish risk ethicist Sven Ove Hansson (2009). Hansson has argued that there is a significant ethical difference between so-called "type-1" errors and "type-2" errors, i.e. false positives versus false negatives. In scientific research, we aim to prevent false positives, that is, false claims that something is the case. But in the context of, for example, policy making about health risks, we want to avoid false negatives, that is, false claims that people are safe while they are at risk. This is based on an ethical consideration, namely that it is prudent to be cautious when dealing with health effects for human beings, in other words that we would rather be safe than sorry. In the context of a pandemic, this can support following the precautionary principle: we don't know how the contagion curve will play out, so it is better to intervene early and be extremely cautious, rather than reach a point where it's too late to prevent disastrous consequences.

For example, the Dutch RIVM had to adjust previous information: initially, it said that Sars-Cov-19 would not be a dangerous virus, not much more than the normal flu. Furthermore, the RIVM initially stated that Sars-Cov-19 could only be spread by people who have symptoms, and therefore not by children as they hardly get sick from COVID-19; they furthermore maintained that aerosols do not play an important role in the spreading of this virus, and that face masks are unnecessary. We now know that all of these claims are wrong. Interestingly, the RIVM and OMT maintained these claims for many months, even in the light of countervailing evidence from other countries, as well as the WHO. Presumably, the OMT and RIVM held on to very high scientific standards concerning sufficient evidence about these matters: as long as it is not entirely clear that these hypotheses are true, they are rejected in order to avoid false positives. As was discussed above, while these are important standards in the context of scientific research, they may not be suitable in the context of public health measures where prevention can also be an important concern, and these standards may not be responsive enough in an urgent crisis. In the words of World Health Organization health emergencies programme executive director Dr. Mike Ryan from 14 March 2020 [sic]:

> Perfection is the enemy of the good when it comes to emergency management. Speed trumps perfection, and the problem in society we have at the moment is everyone is afraid of making a mistake – everyone is afraid of the consequence of error. But the greatest error is not to move. The greatest error is to be paralyzed by the fear of failure (WHO, 2020).

A precautionary approach could have involved a 'what if' exercise at an earlier stage: let's assume the worst and start planning how to deal with this situation. Communicating the difficulties of dealing with uncertain information and complex developments can also help, much more than downplaying these difficulties and claiming certainty. Downplaying uncertainty can easily backfire when things turn out differently (cf. Van Asselt & Vos, 2006), as this will lead to distrust. Striving for certainty cannot always be a priority in situations that are intrinsically uncertain and highly complex, while stakes are high and urgent decision making is needed. As I will discuss in Sect. 11.5, emotions such as compassion and care can highlight important ethical considerations, such as precaution. But first in the following section I will zoom in in more detail on the value-ladenness of COVID-19 policy measures.

11.4   The Value-Ladenness of COVID-19 Policy Measures

The previous discussion highlights that decision-making about possible COVID-19 precautions and their intended positive and negative effects requires ethical reflection. I will now zoom in in more detail on various policy options, by highlighting that they involve important values that need to be deliberated on. In Sect. 11.5 I will then argue that emotions can play an important role in such a deliberation.

In schematic terms, the following COVID-19 strategies can be distinguished. Each comes with underlying assumptions about values and ethical implications:

- Laissez faire: herd immunity

- Business as usual for everyone, but substantial health risks, especially for vulnerable people.
- (Partial) lockdown:
- Everyone affected in terms of secondary health effects and limitation of civil liberties, partial containment of the virus, uncertain evidence about how the virus spreads.
- Containment: intensive testing and selective quarantine:

- Containment of virus, low number of deaths and shorter lockdown, but sophisticated testing, monitoring, and health infrastructure needed; civil liberties restrained.

In the early stages of the pandemic, different countries chose various strategies. They also switched or mixed aspects of these strategies, depending on developments of the pandemic as well as on other societal factors. This is because the development of the pandemic does not just rely on virological issues; it also depends on socio-political and behavioural issues. This means that it requires insight of impacts of behaviour and strategies, as well as reflection on the significant ethical implications of these strategies, by explicating underlying values.

It is crucial to explicitly face the question of how to evaluate different scenarios on how to respond to the COVID-19 crisis. This involves considering available alternatives and their respective advantages and disadvantages, each of which requires ethical reflection. Which values are at stake? How can we assess, compare, and weigh them? Values such as the inconvenience and drastic consequences of social distancing must be weighed against values such as protecting public health and containing a pandemic promptly. This relates to the ethical question as to how to balance direct versus indirect health effects. An example of direct health effects is the need to protect people who are vulnerable to infections. An example of indirect health effects is the need to mitigate the consequences of lockdowns for those who are disproportionally vulnerable to them (e.g. children, young people, people working in the hospitality and cultural sectors).

In any case, those most vulnerable in society will be disproportionally most exposed to the risks of a pandemic. People without health insurance, a steady income, and proper housing are more exposed, for example. These people have fewer means available to ameliorate the impacts of exposure to the virus or lock-down measures than wealthy people and citizens of affluent countries with well-functioning and accessible public health services. The social disruption of a lockdown, for example, is much more profound in a society that does not have a robust social safety net. In a society with such facilities, people whose jobs are at risk due to social distancing policies have better protection, and society's implications will be less disastrous. Finally, the existing political and socio-economic infra-structure in a society is based on ethical considerations. These contextual features need to be considered when ethically evaluating scenarios on how to deal with this and future pandemics.

Furthermore, is increasing herd immunity ethically acceptable if it means that some people will end up in intensive care units when they would not have gotten sick under stricter measures? Given the (specifically in the early stages of the pan-demic) uncertain knowledge surrounding the possible immunity against the SARS-Cov-2 virus, can it be ethically defended to make such an assumption? How to deal with 'triage', that is, how to compare the need for ICU treatment of different patients? In the Netherlands, Covid-19 patients with urgent health care needs are prioritized above other patients who are waiting for non-emergency surgery, and COVID-19 patients stay significantly longer in ICU units than other patients. This means that increasing hospitalization of COVID-19 patients has significant health effects for people with other, less urgent but also eventually life-threatening conditions.

Fundamental and difficult, if not impossible to answer, ethical questions such as 'what is the value of a human life?' are at stake here. Do we opt for a consequential-ist approach to assign a monetary value to human lives, while also allowing human lives to be traded off against each other and other monetized considerations? Or are human lives of intrinsic value, meaning that they cannot be put into a simple equa-tion? The latter seems to be a rhetorical question. From an emotional point of view, we experience the life of a loved one, for example, as infinitely valuable. Deontological approaches in ethics seem to fit better with this insight because they say that we should not use people merely as a means. On the other hand, it is evident that government policies need to balance deontological and consequentialist considerations.

Furthermore, there are ethical questions related to civil liberties, freedom of choice and privacy. Several countries have adopted measures requiring COVID-19 passes, showing that citizens are 'safe' if they have been: (1) fully vaccinated, (2) recovered from COVID-19, or (3) recently tested negative. The privacy of citizens is preserved to a significant degree by not needing to show which of the three crite-ria they meet. Others see this as a disproportionally restrictive measure, which is unsuited for a liberal society. However, respecting people's freedom of choice here comes at the price that vulnerable people – e.g. those who have immune deficiencies and therefore have no or limited choice – are less protected. Even though freedom of choice is of vital importance in a liberal society, we always have to make conces-sions and trade-offs between individual liberties and societal concerns, i.e. the liberty, safety, and health of other people. Safety measures may provide a reasonable trade-off, such as wearing masks and COVID passes.

A crucial issue is that between distributive and procedural justice. While from a distributive justice perspective, it seems fair to have such preventive measures, from a procedural point of view it is essential to have fair means of decision-making that are also experienced as such. Politicians could convey more explicitly that in their decisions, they also take into count ethical and societal concerns and engage with the views of societal stakeholders before making – often complicated – decisions that require trading off or balancing important values. Doing this may have the effect of showing the public the complex value trade-offs that are required in a public health crisis. Such transparency can help the public to see that the decision making was complicated but fair. Making the difficult moral dilemmas and trade-offs explicit as well as reminding people of the responsibilities towards others that come with individual liberties, and appealing to solidarity, can make an important contribution in public deliberation and support for measures. However, these important moral arguments have only rarely been made explicitly by politicians in the Netherlands and other countries.

11.5   The Importance of Emotions

In the previous sections, I have discussed the ethical intricacies of risk assessments and the value-ladenness of COVID-policy measures. The technocratic approach that politicians have primarily used does not suffice to address these issues, but neither would a populist approach be a solution. As mentioned above, populist approaches merely follow the dominant view in society at a particular moment, rather than explicitly addressing the concerns and values at stake. This avoids the problematic ethical deliberations that are sorely needed. Emotions such as compassion, feelings of responsibility, and care can help highlight these ethical aspects. More generally, emotions such as sympathy, empathy, and indignation can play an important role in alerting us to ethically relevant aspects of risks. These and other emotions are at stake within the public at large. Addressing these emotions in explicit ethical deliberation would mean that ethical concerns of the public could be seriously addressed, rather than waved off as in the technocratic approach, or superficially followed without further reflection as in the populist approach, leading to wavering and inconsistent policies.

When thinking about COVID-19 measures, the emotions of diverse stakeholders could therefore play an essential role in highlighting ethical issues and doing justice to important values. The emotions in society can be an essential source of moral insights; indeed, some of the ethical considerations I have mentioned have also been raised by concerned citizens. But, of course, emotions can also be misleading, like all our sources of insight. Emotions can unnecessarily inflate risks, letting them appear overly frightening. At the same time, emotions can make us overlook latent dangers. Intense emotions can magnify our own suffering and thereby ignore the suffering of others (Steinert & Roeser, 2020). This means that emotions must be critically assessed based on scientific information and ethical reflection. But emotions themselves can also play an important role in the latter. Elsewhere, I call this 'emotional deliberation' (Roeser & Pesch, 2016, Roeser, 2018). Furthermore, people's emotions and ethical evaluations can diverge. But rather than eschewing deliberation about these emotions and values, we should engage with them. Such diverging emotions and values can highlight different horns of the complex dilemmas we face. For example, should we require vaccination, e.g. for people working in the care professions or teaching settings or even for all occupations where people interact with each other? Or should we respect people's free choice? But what if that comes at a high price for other members of society, such as people with immune conditions, or those whose medical treatment is getting postponed because of overfull hospitals, or because, say, of secondary health effects due to lockdowns? These are intricate ethical dilemmas, and different stakeholders in society have different views on the best ways to address these. There are, by definition, no easy solutions to moral dilemmas. Instead, they require deliberation, exchange of viewpoints, arguments and experiences to hopefully come to solutions that are acceptable to a broad range of stakeholders. For example, emotions such as a sense of responsibility and concern for others can contribute to putting one's suffering in perspective and being open to policy options that can contribute to the well-being of others. Furthermore, compassion can help understand the suffering of an individual victim, which can disappear in a cold, consequentialist calculation. Opening up deliberation to such concerns can also overcome seemingly unavoidable trade-offs and open new perspectives, by learning from each other and encouraging creative solutions. This can help devise innovative strategies that do justice to public health, economic resilience, and an ecologically sustainable society at the same time.

One might worry that including ethical deliberation and emotional concerns would delay decision making when quick responses are needed. However, there can be explicit ethical deliberation under time pressure, as well as more extensive ethical deliberation, involving stakeholders etc., when preparing strategies in advance. Emotions can actually contribute to a sense of urgency. I will discuss these issues in more detail in the following section.

11.6 COVID-19 Risks, Imagination and Feeling a Sense of Urgency

Emotions, such as a sense of responsibility and empathy, can encourage us to imagine the implications of alternative action options. Works of art and documentaries can facilitate this and contribute to a sense of urgency that currently seems to be lacking in policy approaches to the COVID-19 pandemic, such as in the Netherlands.

It is striking that aside from the warnings of virologists, artists have also warned of the real possibilities of a severe pandemic. There are many examples of this, but

John Suits' 2016 film, *Pandemic*, maps this out in depth. Despite these warnings from the arts and sciences, policy makers around the globe seem to be continuously improvising about how to respond to the current pandemic. One would imagine that every government should have a range of scenarios available with concrete and adaptive plans for a situation like this. Furthermore, on the level of international politics, strategies for dealing with such a situation should have been prepared in advance, e.g. in the context of the WHO. But presumably, other issues were perceived as more urgent, and as we have seen, the Dutch authorities have frequently put recommendations from the WHO concerning COVID-19 aside.

This happened at various stages of the pandemic, not only in the beginning but for example also in the early autumn 2020, when infection rates in the Netherlands started to increase, presumably due to international travel during summer vacations, schools reopening, and seasonal effects. While some other countries were already taking more precautionary measures, the Netherlands waited until the numbers were so high that harsh lockdown measures were eventually unavoidable, in the meantime implying hospitalizations, severe illness and death as well as overburdening the health sector, which had been trimmed down over the last decades in the light of efficiency considerations. Dutch ICU patients had to be admitted to German hospitals at several stages due to Germany's significantly larger ICU capacity. A lack of preparedness also surrounded the early stages of the Dutch vaccination strategy. In autumn 2020, the world was getting unexpected good news: much earlier than hoped for, several vaccines proved to be effective and could be available on a large scale within a few months. Dutch newsreaders could learn on a daily basis how the UK, Israel, Germany, and other countries were preparing their vaccination strategies. They didn't hear much about the Dutch strategies until late December 2020. It was then announced that the vaccination would start in early January 2021, weeks later than the countries mentioned above. The Dutch health minister, Hugo de Jonge, justified this by saying that the Dutch needed more time because they would do things 'thoroughly', implying a less thorough approach by the other countries. Yet, the first weeks of vaccinating were dominated by news about chaotic and inefficient bureaucracy, and multiple changes in strategy, while other countries were making quick progress. Eventually, the Netherlands caught up and now has one of the highest vaccination rates in the world. But a lot of time seems to have been wasted in the early weeks and months. More timely preparations could have alleviated COVID-19 numbers and accompanying direct and indirect health burdens, not to mention the sense of despair that many people felt during that time.

At the time of writing this article (autumn 2021) we see a similar situation in the Netherlands as 1 year before. Despite high vaccination rates, infection rates are rapidly rising, presumably due to the much more infectious Delta-variant in combination with seasonal effects and the loosening of measures when infections rates still were low. While there is a lot of societal concern about this, politicians are slowly and only hesitantly responding, again primarily relying on STEM-based advice by the modellers of the RIVM and the medical and virological experts of the OMT, without consulting social scientists and ethicists. This is despite the fact that there is a growing public tension, for example, an opposition between those who are vaccinated and those who are not, and different views in societies on whether more safety measures are needed or not. This is an issue for which the expertise of social scientists and ethicists could be of crucial importance. Yet, these experts are not systematically consulted by the government or policymakers, except consultation of ethicists for specific medical policy advice concerning vaccination or triage in hospitals and concerning some issues, behavioural scientists at the RIVM and OMT. But, as argued above, all aspects of COVID-policies have significant ethical and societal dimensions, thereby requiring systematically involving the expertise of ethicists and social scientists concerning the overall policy measures.

Emotionally charged human capacities, such as imagination, can play an important role in experiencing urgency, as well as in moral deliberation and in developing and thinking about future scenarios. As mentioned above, the work of artists, filmmakers, and writers can play an essential role in such future scenario thinking. Artworks can appeal to the imagination, make abstract problems more concrete and facilitate ethical deliberation on the implications of such future scenarios (Roeser, 2018). Artworks such as (science fiction) novels and films in which the consequences of a pandemic are described can appeal to the imagination, make abstract problems tangible and thereby facilitate ethical deliberation about the implications of such future scenarios. If policy makers can heed the warnings of artists (such as *Pandemic*, for example), then the arts may have potential to help catalyse future pandemic-prevention strategies, taking into account the implications for public health, as well as for the economy and well-being of different population groups.

11.7 Conclusion

The current COVID-19 crisis highlights that decision-making about risks always requires scientific knowledge to be accompanied by societal and ethical considerations. My approach to emotions in the context of risk offers an alternative to the technocratic or populist approaches used to combat COVID-19. Emotions are a rich and valuable resource that is wrongly rejected in decision-making about risk and uncertainty. The current approach should be enriched, focusing on citizens' concerns, involving ethical reflection on different choices and policy options. Emotions such as compassion, feelings of responsibility, and concern can help us reflect on the ethical implications of the difficult decisions we face. In the current situation and coming years, we will need all the sources of insight we have at our disposal to meet the enormous challenges of the COVID-19 crisis as well as possible future pandemics. So indeed, we need to consider the insights of virologists and medical experts. Still, we also need expertise from ethicists, social scientists, and the arts and humanities to take social and ethical considerations into account. In order to take on the severe challenges of this situation, we need to draw on our rich human capacities: scientific knowledge, insights from social sciences, arts and humanities, and emotional capacities. Rather than dismissing emotions, we should embrace them as a vital resource. Emotions such as compassion and feelings of responsibility and care can help us to reflect on the ethical implications of the hard choices we face. They can play an essential role in motivating actions of solidarity and courage that can hopefully contribute to solutions to the ongoing as well as future pandemics.

Acknowledgements I would like to thank Matthew Dennis and the other editors for their very helpful comments on an earlier version of this chapter. The chapter incorporates updated and significantly expanded passages from a Dutch article: Sabine Roeser (2020), 'Corona, risico's en morele emoties', Tijdschrift voor Bioethiek 27:3 pp. 7–9 (reprinted at the blog 'bijnaderinzien'). Available at: https://nvbioethiek.files.wordpress.com/2020/07/podium-20-3-naar-een-morele-agenda-na-de-coronacrisis.pdf

References

Asveld, L., & Roeser, S. (Eds.). (2009). *The ethics of technological risk.* Earthscan.

Damasio, A. R. (1994). *Descartes' error: Emotion, reason and the human brain.* G.P. Putnam.

De Hollander, A. E. M., & Hanemaaijer, A. H. (2003). *Nuchter omgaan met risico's: Milieu—En natuurplanbureau (MNP)—RIVM.* RIVM.

Fisher, M. (2020). *Coronavirus 'hits all the hot buttons' for how we misjudge risk.* https://www.nytimes.com/2020/02/13/world/asia/coronavirus-risk-interpreter.html, visited: 3 November 2021.

Frijda, N. (1986). *The emotions.* Cambridge University Press.

Furtak, R. A. (2018). *Knowing emotions: Truthfulness and recognition in affective experience.* Oxford University Press.

Hansson, S. O. (2009). An agenda for the ethics of risk. In L. Asveld & S. Roeser (Eds.), *The ethics of technological risks* (pp. 11–23). Earthscan.

Hume, D. (1975) [1739-40]. L. A. Selby-Bigge (Ed.), A *treatise of human nature,* 2nd ed. revised by P.H. Nidditch, Clarendon press, Oxford.

Kahneman, D. (2011). *Thinking fast and slow.* Farrar, Straus and Giroux.

Loewenstein, G. F., Weber, E. U., Hsee, C. K., & Welch, N. (2001). Risk as feelings. *Psychological Bulletin, 127*(2), 267–286. https://psycnet.apa.org/doi/10.1037/0033-2909.127.2.267

Moore, G.E. (1988) [1903]. *Principia Ethica.* Promotheus Books.

NOS. (2020). https://nos.nl/artikel/2328889-rond-19-00-uur-wordt-duidelijk-hoe-gaan-we-verder-na-6-april; visited: 3 November 2021.

NOS. (2021a). https://nos.nl/nieuwsuur/collectie/13881/artikel/2403299-reconstructie-nederland-koos-te-laat-voor-bestrijden-corona; visited: 3 November 2021.

NOS. (2021b). https://nos.nl/nieuwsuur/collectie/13881/artikel/2403300-verantwoording-reconstructie-corona-aanpak-nieuwsuur; visited: 3 November 2021.

NOS. (2021c). https://nos.nl/nieuwsuur/collectie/13881/artikel/2403432-gommers-en-kuipers-kritisch-sturen-op-besmettingen-had-reguliere-zorg-minder-belast; visited: 3 November 2021.

NOS. (2021d). https://www.youtube.com/watch?v=Zp0MDXy7F2k; visited: 10 December 2021.

Nussbaum, M. C. (2001). *Upheavals of thought: The intelligence of emotions.* Cambridge University Press.

Prichard, H. A. (1912). Does moral philosophy rest on a mistake? *Mind, 21*(81), 21–37. https://doi.org/10.1093/mind/XXI.81.21

Roberts, R. C. (2003). *Emotions: An essay in aid of moral psychology.* Cambridge University Press.

Roeser, S. (2006). The role of emotions in judging the moral acceptability of risks. *Safety Science, 44*(8), 689–700. https://doi.org/10.1016/j.ssci.2006.02.001

Roeser, S. (2011). *Moral emotions and intuitions.* Palgrave Macmillan.

Roeser, S. (2018). *Risk, technology, and moral emotions*. Routledge.

Roeser, S., Hillerbrand, R., Peterson, M., & Sandin, P. (Eds.). (2012). *Handbook of risk theory: Epistemology, decision theory, ethics, and social implications of risk*. Springer. https://doi.org/10.1007/s11948-012-9362-y

Roeser, S., & Pesch, U. (2016). 'An emotional deliberation approach to risk'. Science, Technology & Human Values, 41, 274-297.

Slovic, P. (2010). *The feeling of risk: New perspectives on risk perception*. Earthscan.

Solomon, R. C. (1993). *The passions: Emotions and the meaning of life*. Hackett Publishing Company.

Steinert, S., & Roeser, S. (2020). Emotions, values and technology: Illuminating the blind spots. *Journal of Responsible Innovation,* 7(3), 298–319. https://doi.org/10.1080/2329946 0.2020.1738024

Sunstein, C. R. (2005). *Laws of fear*. Cambridge University Press.

Sunstein, C.R. (2020). The cognitive bias that makes us panic about coronavirus. *Bloomberg.* https://www.bloombergquint.com/gadfly/coronavirus-panic-caused-by-probability-neglect   ; visited 10 December 2021.

Van Asselt, M., & Vos, E. (2006). The precautionary principle and the uncertainty paradox. *Journal of Risk Research, 9*, 313–336. https://doi.org/10.1080/13669870500175063

WHO.   (2020).   https://www.youtube.com/watch?v=e-XPL91r0uk&t=4s;   visited   26 November 2021.

Open Access This chapter is licensed under the terms of the Creative Commons Attribution 4.0 International License (http://creativecommons.org/licenses/by/4.0/), which permits use, sharing, adaptation, distribution and reproduction in any medium or format, as long as you give appropriate credit to the original author(s) and the source, provide a link to the Creative Commons license and indicate if changes were made.

The images or other third party material in this chapter are included in the chapter's Creative Commons license, unless indicated otherwise in a credit line to the material. If material is not included in the chapter's Creative Commons license and your intended use is not permitted by statutory regulation or exceeds the permitted use, you will need to obtain permission directly from the copyright holder.

Chapter 12
How to Balance Individual and Collective Values After COVID-19? Ethical Reflections on Crowd Management at Dutch Train Stations

Andrej Dameski, Andreas Spahn, Caspar A. S. Pouw, Alessandro Corbetta, Federico Toschi, and Gunter Bombaerts

12.1 Introduction

The COVID-19 pandemic has changed the way individuals have had to behave when in contact with other individuals. SARS-CoV-2 has exhibited a swift and exponential spread rate. Clusters of individuals (families, public events and gatherings, crowds, etc.) have sped up the transmission, and during 2020 they accounted as responsible for 50–80% of all reported cases (Hozhabri et al., 2020). Although the mortality rate of COVID-19 is between 2.0% and 3.8% (Hozhabri et al., 2020; Novel, 2020), which is significantly lower than previous coronavirus epidemics (e.g. SARS with ~10% and MERS with ~35%), it is still relatively high nonetheless, especially among vulnerable parts of the population: elderly, people with chronic diseases, the immunocompromised, etc.

Faced with these numbers, governments reacted by attempting to control the spread of the virus through imposing measures (e.g. social distancing, minimisation of crowds and public gatherings, mandating the wear of face masks, and others) which were assumed to be capable of lowering the transmission rate.

Virtually all these measures have asserted that collective values (should) have primacy in times of crises, in particular during the COVID-19 pandemic, and that those restrictions of individual freedoms are thus an acceptable response. Nevertheless, these measures were only temporarily successful in stopping the spread of the virus among the population, since the pandemic has resulted in several 'waves' of infections and in multiple mutations (strains), some of which have proved to be even more contagious (e.g. Delta, Omicron).

A. Dameski (✉) · A. Spahn · C. A. S. Pouw · A. Corbetta · F. Toschi · G. Bombaerts
Eindhoven University of Technology, Eindhoven, Netherlands
e-mail: a.dameski@tue.nl; a.Spahn@tue.nl; c.a.s.pouw@tue.nl; a.corbetta@tue.nl; f.toschi@tue.nl; g.bombaerts@tue.nl

© The Author(s) 2022
M. J. Dennis et al. (eds.), *Values for a Post-Pandemic Future*, Philosophy of Engineering and Technology 40, https://doi.org/10.1007/978-3-031-08424-9_12

However, this (primarily governmental) assertion towards the primacy of collective values in crises (which at times is described as emerging '(latent) authoritarianism'; see, for example, Hoxhaj and Zhilla (2021) and Thomson and Ip (2020)) has received pushback after initial public support. This, paired with the perceived general ineffectiveness (Asongu et al., 2020) and perceived needlessness or arbitrariness of the measures, and with the secondary toll of the imposed containment measures (e.g. rising depression, anxiety and suicide rates, loss of employment, traumatised interactions between people; see Sikali (2020) and Beeckman et al. (2020)), has motivated a more profound public debate on the balance between individual and collective values, and whether some belonging to one or the other group have priority in times of crises. We explore this debate in Sect. 12.2.

The pandemic has also given rise to questions on how to effectively control the behaviour of groups of people and how to use technologies for pandemic crowd control. This has added a new dimension to the discussion of crowd control technologies. For this purpose, we conduct an exploratory ethical analysis of recent sociophysics research findings from Pouw et al. (2020), which is focused on monitoring crowds on train stations. This case study is part of a research project aimed at understanding the movement of individuals and crowds within train stations to help better manage the flow of travellers e.g. in peak moments. In 2021 a research consortium was created to also include societal aspects of crowd-management with the help of psychological and ethical research.

In this chapter we present an exploratory ethical analysis of recent findings of this empirical project during the COVID-19 pandemic. The aim of this exploratory study is to identify both key research questions and initial findings that will be relevant for the further ethical investigation of collective and individual agency and the balancing of individual and collective values in crowd nudging.

We identify three important research questions for philosophy and ethics of technology that require interdisciplinary cooperation between empirical and philosophical research and present initial reflections on each of these three questions: How can we understand and conceptualise the relation between collective and individual values and agency (Sect. 12.2.1)? How should we balance individual and collective values post-COVID (Sect. 12.2.2.)? What role can and should crowd management technology play in this balancing acts (Sect. 12.2.3)? In the next section we elaborate these three aspects from a philosophical perspective, before applying these three questions to the case of crowd management at train stations., which we explore in greater detail in Sect. 12.3. below.

12.2   The Normative Background of the Current Pandemic: Collective Versus Individual Values in Times of Crises

To understand collective and individual values and their balance, we begin with a brief discussion on what values are. For our chapter, we follow the definition of values of Schwartz, developed in his *theory of basic values*. According to Schwartz, values are "trans-situational goals, varying in importance, which serve as guiding principles in the life of a person or group" (Schwartz, 1992, 1994, 2017). Schwartz identifies 19 distinct values within 12 value factors (clusters) in this latest iteration of his theory. In addition, he identifies the properties that values must have and the dynamic functions they need to fulfil. In Schwartz's words, values "should be grounded in one or more of three universal requirements of human existence with which people must cope: needs of individuals as biological organisms, requisites of coordinated social interaction, and survival and welfare needs of groups"[1] (Schwartz, 2017).

As we can see, Schwartz keenly recognises that values pertain both to individuals, groups and social interactions. With this in mind, we can take as:

- *individual values* those that predominantly pertain to individuals;
- as *collective values* that predominantly pertain to groups.

In regards to the COVID-19 pandemic, the values that will be of most interest here are:

- Collective values: collective safety, collective responsibility, conformism;
- Individual values: individual autonomy, freedom, safety, responsibility, and privacy.

We will use these more commonly used terms further down.

12.2.1 Is There Such Thing as Collective Values (and Collective Entities)?

We have discussed values and their split into two groups of individual and collective. However, the COVID-19 pandemic has shown that it is not clear who should take responsibility for realising these values. This is especially important for crowd management and the use of technology, where the following question arises: is there such a thing as collective agency, intentions and responsibility of crowds; aside, and next to, the agency, intentions and responsibility of individuals?

In other words, our driving question is whether collectives are something different from the simple set of individuals that comprise them, and whether this means that collective values can also, or only, apply to collectives *per se*.

This is a centuries-long ongoing debate between sceptics and proponents, especially regarding collective responsibility (Smiley, 2017). More recently, methodological individualists oppose ascribing responsibility to groups and collectives *per se*, and may only subscribe to a 'collective' responsibility as a distributive

---

[1] Additionally, values should: "... (1) focus on attaining personal or social outcomes, (2) express openness to change or conservation of the status quo or (3) serve self-interests or transcendence of self-interests in the service of others, and (4) promote growth and self-expansion or protect against anxiety and threat to self" (Schwartz, 2017).

phenomenon, i.e. distinct individual responsibility distributed to each member of a collective. For some sceptics, groups cannot form selfhood and thus lack intentions, wills, agency, and actions. For those with this view, all the above can only be individual and therefore consist only of discreet phenomena.

Nevertheless, proponents of shared intentions (collective intentionality), shared agency, and collective responsibility assert that collectives can indeed have intentions, will(s), agency, and actions, which cannot be explained away as purely individual (distributive) phenomena (Schweikard & Schmid, 2021; Smiley, 2017). Such phenomena might be tradition or shared practices, patriotism, sense of belonging to a group (e.g. one's own family), feelings of societal pride or shame, conformism, organisational identity, life, and activities (such as corporations, states, organisations).

In other words, these are non-distributive phenomena, at least sometimes caused by 'collective intentional agents' (Corlett, 2001, p. 575; in Smiley, 2017). However, proponents of collective values and agency differ on the properties these collective entities have, as well as the criteria that determine whether a set of individuals has indeed integrated enough to take on a collective 'intentionality' of a sort – and thus be a proper 'target' for collective responsibility, accountability, and liability.

Systems theory may come to the rescue here, especially its notions of integration and emergence. The classical approach to defining a system is that a system is a set of things and relations between those things (see Klir, 2013). Without any of the two sets (things and relations), there cannot be a system. Inversely, once a particular set of things 'acquire' a set of relations between themselves, these things have been integrated into a system, and thus the system *emerged.*

Integration into a collective does not necessarily imply that its components (the individuals) lose all independence and personal agency, and thus are rendered into mindless automatons. On the contrary, personal agency and autonomy can remain and motivate individuals to individual action, not always complying with the collective.[2] Additionally, integration in a collective does imply, at least in some cases, a temporal emergence of collective intentions (see Schweikard & Schmid, 2021). Some authors might even consider that, besides obtaining collective intentions, individuals can sometimes integrate with/in collective entities (see, for example, Durkheim in LibreTexts, 2020).

Therefore, we argue that some values, such as responsibility, safety, autonomy, privacy and others, can apply to entities that are integrated in a way that becomes more than just the sum of its parts (individuals), both from the less controversial notion of collective intentions and from the more controversial integration in a collective entity. The individual notions of these values continue to apply to individuals in parallel.

This brings us to our first set of research questions that we want to apply to the case study: how can individual and collective agency be understood in empirical

---

[2] Just like how one can identify and act as a member of a particular ethnic community, while also identifying and acting as an individual (see Borch, 2009; Johanssen, 2016 and Schweikard & Schmid, 2021).

research? How are individual and collective agency conceptualised in the empirical research study? (see Sect. 12.3.)

12.2.2 The Balance Between Collective and Individual Values, and the Impact of COVID on This Balance

In the ethical and – even more importantly – legal spheres of liberal democratic societies,[3] there can be vigorous attempts to discover the proper balance between individual and collective values. By 'balance' here, we mean a manner of applying values to decision-making, in particular situations, as an attempt to maximise the flourishing of both individuals and collectives (including societies at large) side-by-side, while attempting to minimise trade-offs in achieving this goal.[4] Another way of understanding this is as an attempt to maximise the satisfaction of individual and collective values (for example, the ones listed by Schwarz) in conditions of particular situations and limited resources.

This is an age-old inquiry. The evolution of thought in Europe has been moving from the absolute primacy of the collective (in primordial human tribes) to a slow emancipation of the individual. In Europe, this was bootstrapped by Christianity and its assertion of the divinity of the human being. It resulted in reversing the dominance of the collective, and the primacy of the individual over the collective was born (Hösle, 2004; Maine, 2007; Triandis, 1995).

Legal and moral systems were slow to adopt this change, especially as there was a reactionary pushback towards the reassertion of the importance of the collectives with the rise of the nation-states and dominantly-collectivist thought (anarcho-communism, communism and socialism, fascism, national-socialism, and communitarianism). Finally, the most recent revolutionary development is the creation of the United Nations and its founding documents that pertain to universal human rights and freedoms (Ishay, 2020; Donnelly, 2013).

Most of the developed democracies today are a somewhat balanced mixture between collectivism and individualism. Perhaps this is unsurprising because sociological research shows that human beings are disposed to having (1) genetically inherent types of collectivist and individualist instincts[5] and (2) a culture-gene coevolutionary coupling process (Chiao & Blizinsky, 2010; also see Haidt, 2012).

---

[3] We understand such societies as those which regard (1) individuals, (2) the rule of law, and (3) majority voting as vital structuring principles.

[4] Or, in simple words, we may interpret this as treating each individual as equal in inherent value to every other individual, and to every collective—and vice versa. Therefore, if each individual and/ or collective is equally important, the values that pertain to them are equally important and should get equal attention. This is, of course, the ideal state of matters that might not actually take place in practice.

[5] Albeit set at differing ratios across peoples and cultures (Way & Lieberman, 2010).

The foundational UN documents (see United Nations, 1945, 1949) affirm the existence of both individuals and collective entities (nations, peoples, families) and affirm both individual and collective rights and freedoms. By doing this, they attempt to validate both individual and collective values, even if they strongly emphasise the rights of the individual by their very nature (Spahn, 2018).

The general way relatively stable European democratic societies have gone about designing their legal systems and institutions has been to develop human rights frameworks that specify individual and collective rights (and therefore assert both types of values), and also to assert the primacy of particular individual or collective values in particular contexts.

For example, governments are generally tasked with promoting the welfare of individuals, collectives and broader society; with special focus on individuals, in order to ensure they are not being dominated by the other two. Nevertheless, in times of emergency (e.g. wars, terrorism, epi/pandemics, vis major and major tragedies), governments often exercise emergency powers that can temporarily suppress individual values, rights, and freedoms, with the purpose of protecting the welfare of collectives and society at large.

There are also some highly specific areas where the government is allowed to assert the primacy of the collective over individuals even before a crisis occurs. This is often done to prevent a crisis. Examples include taxation, mandatory pension contributions, public health participation, law and order services, security regarding critical domains and technologies (nuclear, military, cyberspace, etc.).

In these highly specific mandated cases, there are usually strict boundaries in place to prevent abuse of the state apparatus over individual values, rights, and freedoms. Suppose there is an attempt to circumvent or transgress these boundaries by the government and the state apparatus. In that case, this is a shift towards authoritarianism and a turn away from a healthy democratic process.

Stable democratic societies, therefore, do recognise that both individual and collective values exist, that they are equally important, and that both should receive due attention and affirmation. However, they also realise that particular values of the two types can conflict with each other, especially at times of crisis. When this happens, a discrete balancing solution ought to be discovered through public discourse and decision-making that includes all affected stakeholders. These discrete balance points can favour one or the other types of value in particular contexts.

Finally, the aggregation of all these balancing points, along with the general balancing principles between the two types, comprise the general societal value balance between collective and individual values for each particular society.

As we mentioned in the introduction, the COVID-19 pandemic has resulted in a strong – if temporary – assertion of the primacy of collective values (collective safety, health, conformism, and responsibility) over individual values (individual autonomy, privacy, and responsibility). The assumed rationale is the need to solve the crisis in a predominantly collectivist fashion, which is assumed to be more efficient than in a predominantly individualist or hybrid one. This seems to indicate that support for (quasi-)authoritarian approaches appear to increase in times of perceived threat and crises (Feldman & Stenner, 1997).

Faced with the impending pandemic, governments have taken their leeway to employ measures compliant with this assertion in varying degrees (see, for example, Amer et al., 2021). However, at times and in particular national contexts, this has turned into stringent infection-containment measures that severely disrupted individual rights and freedoms.

For example, China has initially reacted to the spread of the SARS-CoV-2 virus by applying a strict 76-day lockdown and curfew on Wuhan (BBC, 2021). Almost all countries have imposed temporary bans on international travel, internal curfews, mandatory social distancing and hygiene measures, and vaccination status proofs to be able to access specific spaces. Some otherwise democratic countries which explicitly put a strong focus on individual human rights and freedoms, such as the USA (Kimball and Josephs, 2021), Italy, Austria (The Guardian, 2021) and Australia (Al Jazeera, 2021), have imposed almost draconian and potentially discriminatory measures. These include strict curfews lasting many months, mandatory vaccination proofs in order to be able to work in private businesses (or even enter shops and buy food), and lockdowns that apply based on vaccination status.

After initial support for the measures that national governments have implemented to contain virus spread, citizens have started pushing back against these measures through passive and active means.[6] Additionally, the level of adherence to mandated measures or recommendations seems to be also connected to the perceived severity of the risk of infection. This has, for example, been observed among Danish students. The healthier and younger they considered themselves, the less were they concerned with getting infected, and the less they adhered to the measures or recommendations (Berg-Beckhoff et al., 2021).

This pushback, we contend, can be interpreted as re-assertion of the importance of individual values, disapproval of the governmental shift of balance towards collective values, and a demand to restore the balance to a pre-pandemic (or to another more balanced) position.

Due to the recent increase in COVID-19 infections, many governments continue to assert that collective values have a primacy during this crisis and that they have the authority to mandate such measures. One 'silent' portion of the population – assumed to be significant – supports governmental measures to prevent the spread of the virus, while another – notably louder – portion of the population vociferously rejects this assertion (see Keiser, 2021).[7]

---

[6] *Passive means* include decreasing compliance with the imposed containment measures, such as social distancing, lockdowns and curfews, and avoidance of social contacts outside one's 'bubble', and other (we explore empirical findings in this regard below in Chap. 3, by analysing the findings from the sociophysics paper of Pouw et al., 2020; also, see Beeckman et al., 2020). *Active means* include protests, explicit disrespect towards the imposed measures, refusal to vaccinate and provide vaccination status, disrespecting mandatory quarantine, and other. The strength of pushback against governmental measures seems to be strongly connected to the level of trust and confidence in the government to tackle the pandemic, but also modified by factors such as mental health and wellbeing, worries about future adversities, and social isolation and loneliness (Wright et al., 2020).

[7] This latter group includes individualists, libertarians, minarchists, anarchists, vaccine-, governmental-, and Big Pharma sceptics, members of strict religious groups, etc.

Scholars have shown that this type of polarisation seems to be increasing consistently (Jungkunz, 2021; Keiser 2021). It is related to political (Maoz & Zeynep, 2010) and societal instability (Keiser, 2021), as well as increasing distrust in government, the state apparatus and institutions (Jones, 2015). If it continues for a prolonged period and converts into a chronic societal phenomenon, such distrust might also result in a decreased "willingness to obey laws" (Jones, 2015).

This brings us to our second set of research questions for ethics of technology: what can we learn empirically about the willingness of individuals to obey rules that prioritize the common good in times of crises? Under which conditions are individuals more or less likely to behave in societally desirable ways?

12.2.3   The Use of Crowd Management Technology Pre- and Throughout the Pandemic

The balancing between individual and collective values is not only a question for policy, but also plays a role in the design and usage of technology. Increasingly technologies play a significant role in steering the behaviour of both individuals and groups. As a result, there is a growing potential for technologies to monitor and influence human actions. This has been discussed in the ethics of technology under various labels, most prominently as so-called persuasive or behaviour change technologies (Fogg, 2002; Spahn, 2012).

With increasing digitalisation, technologies can quickly take over the role of nudging people, as developments in ICT allow to monitor the behaviour of individuals or groups of people, collect increasing amounts of data about users and inform, nudge or persuade people to change their behaviour at just the right time (Spahn, 2020). Individual users can download, for instance, e-coaching apps that help them lose weight or COVID tracking apps that inform them about risk encounters. At the same time, digital technologies can be used to monitor and steer the behaviour of large crowds, for instance, attempting to direct traffic flows in cities or crowd management at train stations.

Thaler and Sunstein (2008) have advocated the use of 'nudges' to help people act in line with their values. They propose a framework of so-called 'libertarian paternalism'. Since the environment we live in influences our choices and behaviour, designers of technology can use this to their advantage and push people to behave differently. Thaler and Sunstein argue that these interventions should be in line with the values that people themselves embrace (hence the paternalism part of the label), while at the same time, they should leave people the freedom to override or ignore these nudges (hence the libertarian element of their view).

However, the experience of the pandemic points to a shift in the usage and debate about these nudges. There is a rich literature on the question of whether it is ethical to nudge people since this seems to be interfering with their autonomy and freedom of choice (e.g. Engelen & Nys, 2020; Hausman & Welch, 2010; John et al., 2013), especially in cases in which individuals might not share the values of the technologies or the designers of these technologies.

Therefore, the question of influencing group behaviour to cultivate social values implies an analogous difference between individual and collective nudges. The original idea defended by Thaler and Sunstein was that individuals could accept nudges in line with their values (such as health, wealth and happiness). Nevertheless, nudges can also be used to influence individuals and crowds towards behaviour or values that are seen to be in line with the greater good, even though the individual might not embrace them. This might be the case in, for example, sustainability (Schubert, 2017), general health care (Capasso & Umbrello, 2021), or in adherence to COVID-19 rules.

This brings us to our final set of research questions for ethics of technology: how can we use technologies to influence individual behaviour? How willing are people to accept nudges that prioritise collective values?

12.3 Crowd Control – a Case Study from Sociophysics

We now move to focus on one particular crowd control technology, developed and used by the SRCrowd project of the Physics of Social Systems group at the TU/e (Eindhoven University of Technology), and described in a recent sociophysics paper by Pouw et al. (2020). This technology was and is used to analyse crowd behaviour at the Utrecht train station, the Netherlands, before and during the COVID-19 pandemic.

We use our exposition of the three relevant domains of philosophical research questions to structure our explorative analysis of the case study, and to identify important steps for future research.

12.3.1 Individual and Collective Agency

To conduct their studies, the researchers had to find ways to empirically identify different types of collectives and the relations between collective behaviour and individual agency. The researchers have managed to determine a variety of crowd phenomena and properties, such as *family group relationships*, *offenders*, *repeated offenders*, *crowd density*, (potential COVID-19) *exposure time*, *relevant interactions*, *family-groups subtransitive closure*, *total individual exposure time*, *pairwise exposure time* and *distance*, and *evolution of behaviour* before and throughout the COVID-19 pandemic.

It is worth noting that all these phenomena and properties (except evolution of behaviour) can be tracked and determined in real-time. For example, this means that crowd behaviour can be tracked and analysed by using live feed from trackers around train stations or later applied to such data that was pre-gathered.

For the relation between individual agency and collectives, the identification of group relation is an interesting way to identify family-group relation based on observational data. The criteria set for the predetermined amount and distance for two or more people to be considered a family group are people who have a pairwise distance of fewer than 1.5 m for 90% per cent of the time and are within 1 m for 40% per cent of the time (Pouw et al., 2020). The rationale is that "pedestrians who followed the same trajectory, thereby being in mutual proximity for the major part of their persistence time, and who are comfortable for extended periods in each other's private space ($r \leq 1$ m) most likely belong to the same family-group" (Pouw et al., 2020, p. 8).

The research also attempts to define and identify **unwanted collective behaviour**.

The primary measure used for this purpose is the so-called '*Corona event*', where "two people, not belonging to the same family, get closer than a threshold distance D" (Pouw et al., 2020, p. 2). The distance is defined as equal to or less than 2.5 m. This criterion is modified further in the paper by using a particular minimum contact time of 0 s, 5 s, and 30 seconds (Pouw et al., 2020). Travellers that act irresponsibly and don't respect these conditions (i.e. by triggering a Corona event) are labelled *offenders*, and travellers that repeatedly disrespect the conditions are labelled *repeated offenders*.

Combining the operationalisation of collective units, such as e.g. families, and unwanted behaviour, such as a corona event, allowed the researchers to successfully discern family groups and offenders of social distancing measures. For example, Fig. 12.1 above describes the acceptable behaviour of a 'family group (a) and an unacceptable behaviour of a (repeated) offender (b). This shows that sociophysics research can aid sociological, psychological and ethical research, if the limitations of such data-driven analysis are taken in consideration.

12.3.2   Adherence to Rules and the Balance Between Individual and Collective Values

Through tracking and analysing crowd behaviour on the Utrecht platform, especially regarding the so-called Corona events (which we previously took as a proxy for unwanted collective behaviour), the sociophysics research by Pouw et al. (2020) can provide valuable ethical insight into crowd behaviour, and (collective) responsibility.

When trying to manage crowds to adhere to ethical rules, such as the corona measures, it is essential to bear the phenomenon of *rule fatigue* in mind. Regarding the adherence to corona measures, it was found that travellers suffer from what can be defined as 'rule fatigue' i.e. the steadily decreasing adherence to suggested or mandated behaviour-regulating rules over time. Furthermore, the researchers found that as the use of the platform slowly recovered from the initial dip at the beginning of the pandemic (i.e. from weeks 17 to 26), so the average individual exposure time for distances between 0.5 m and 2.5 m increased (i.e. offences and (repeated) offenders statistically increased), thus increasing the risk of infection (Pouw et al., 2020).

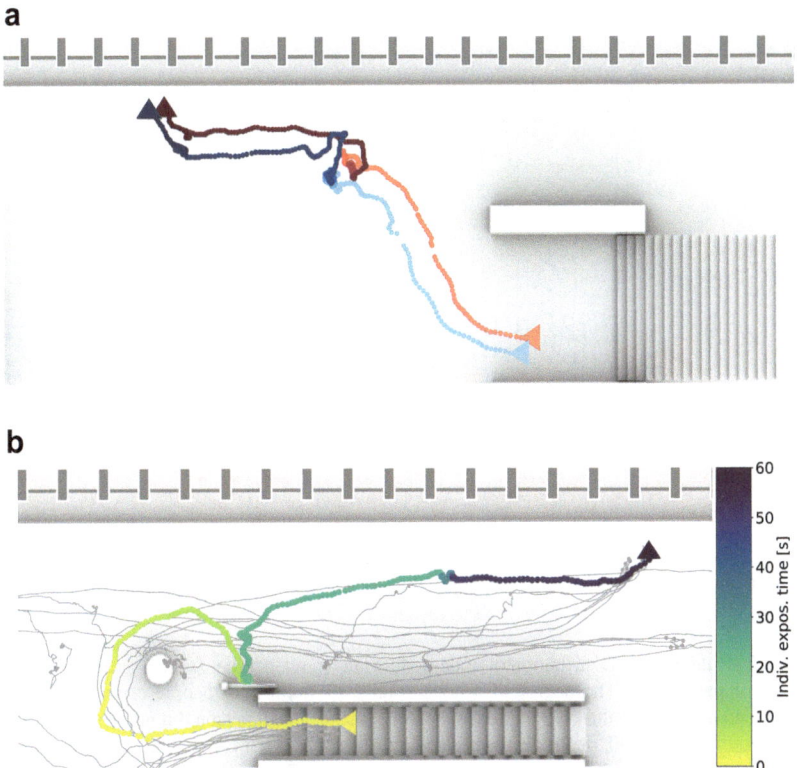

Fig. 12.1 (**a**) Detected clique consisting of two nodes representing two people travelling together. Both entering the platform through the stairs, waiting together for the next train to arrive and finally boarding the train through the same door. The hue of the trajectories is proportional to the time spent on the platform. Lighter hue when the people enter and a darker hue when they leave. Jump in hue, indicating the place where the travellers were waiting. (**b**) Detected node with degree higher than 10, i.e. a repeated offender who violates physical distancing with more than 10 other people

This trend was only temporarily 'reset' when the train schedule was changed on June first, after which the number of repeat offenders started steadily increasing again (see Fig. 12.2. above). On the 1st of June 2020, the train schedule was restored to pre-pandemic levels, which suddenly increased the respect for physical distancing requirements. This change has taken place possibly because the train schedule change has 'shaken out' people out of their habituated abiding of social distancing rules (i.e. behavioural inertia, see below); because it made respecting these rules easier by reducing the load on the platforms; or a combination of these and other factors in play. Similarly, rule fatigue seems to be involved again, since the respect for the social distancing measures again steadily decreases from this date onward. Part of rule fatigue appears to be people developing behavioural inertia as they get used to measures, resulting in adherence in an 'automated' fashion without paying much conscious attention, which might be why compliance decreases over time.

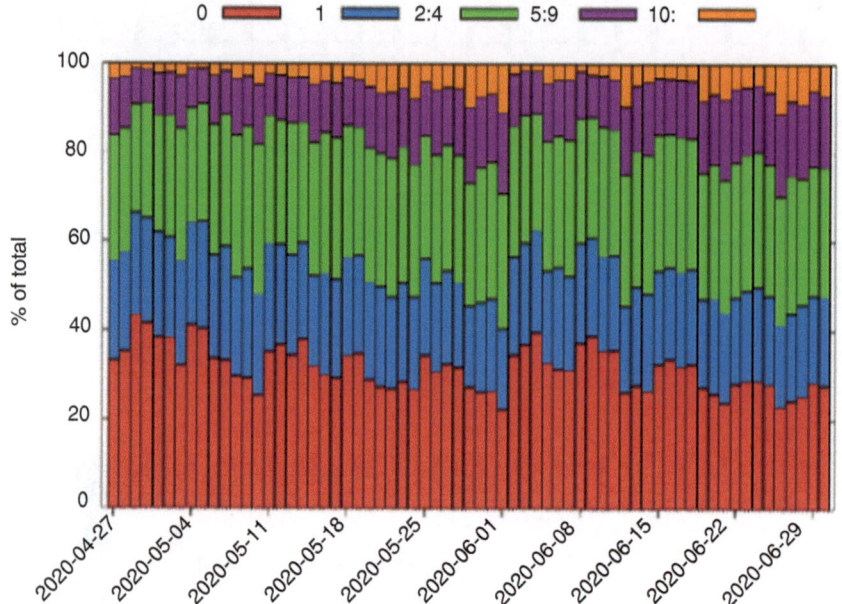

Fig. 12.2 Rule fatigue. Distribution of node-pedestrian degree per day as a percentage of the total number of passengers. The degree of a node counts the number of people encountered with a mutual distance smaller than 1.5 m (hence, degree 0 means that a person did not have any Corona event). Pouw et al. (2020, p. 10) observed that high-degree nodes, i.e. repeated distance offenders, increased steadily until the train schedule changed. The schedule change on June 1st yielded a temporary drop in the offender percentage, after which it started increasing again

Rule fatigue can be considered an ethically relevant phenomenon that can be considered when designing and employing crowd management rules to ensure the best possible effect from enacted rules while not irritating people that are supposed to be following them.

Additionally, in parallel, the average individual exposure time for distances below 0.5 m. remained the same, which might be interpreted as individuals being adamant about keeping their personal distance from strangers.

12.3.3 Acceptance and Acceptability of Social Rules

Another phenomenon that was discovered in the paper, and which is relevant to the ethical question under which circumstances are people willing to follow the social distancing rules, was that travellers find keeping enough distance with unrelated other passengers manageable until the density threshold of 1 pedestrian per 5 m^2 is reached (with minimum contact time threshold of 30 s). After this threshold is passed, the number of Corona events sharply increases.

This result is not only a physical boundary condition, as this density implies that people are on average 2.2 m ($\sqrt{5}$ m^2) from each other. As the authors themselves point out, this "... can suggest an increase in difficulty in following distancing rules around this density level" (Pouw et al., 2020, p. 10). Ethically interpreted findings such as this one can positively inform policymaking in designing better and more ethically acceptable rules for crowd control and social distancing.

A third interesting observation that can help ethically acceptable (e.g. privacy-preserving) crowd tracking while informing policymaking is extracting the statistical average of family groups from the total number of travellers. For example, Pouw et al., by using the criterion of people having a pairwise distance of less than 1.5 m. for 90% of the time and less than 1 m for at least 40% of the time, have managed to identify (the percentage of) family-groups which are allowed to stay close together without infringing upon social distancing rules. On average, around 11% of all visitors of the platform belonged to family groups. Interestingly, this average did not change throughout the analysis even though the number of visitors, density, and offenders did.

Aside from the identified phenomena above on which we put some focus, a closer collaboration might result in a better understanding of a plethora of other crowd phenomena relevant to ethics.

Some examples would be *pairwise exposure time* and *pairwise distance statistics, total individual exposure time* (which might help in determining the risk a particular individual has to become infected); *family subgraph transitive closure* (for identifying people that consider themselves mutually close or intimate); *pedestrian density* and *average pairwise distance* as proxies for what people consider their personal space; and others.

Future research should complement these empirical findings with qualitative insights about the experiences of travellers and their motives for adhering or breaking of social rules, such as the distance keeping. This should give insight into the psychological acceptance of (a) monitoring of behaviour with regard to privacy, (b) of social rules and norms for desired behaviour and (c) nudges to adhere to desired behaviour. These findings can inform the ethical debate on the moral acceptability of crowd nudges and the right balancing point of individual and collective values in a post-COVID-19 world.

12.4   Conclusion – The Future of Crowd-Management and the Relations Between Individual and Collective Values in a Post-COVID-19 World

We initially stressed the importance of balancing individual and collective values and how an emergency such as the COVID-19 pandemic can potentially shift this balance. Then, we focused on a crowd management research case that was held at the Utrecht train station and found several examples on how social physics and ethics research can mutually support each other.

Finally, we want to discuss three important lessons learned from the COVID-19 crisis: (1) the importance of the empirical evidence for the individual-collective debate, (2) the ethics of individual and collective nudging, and (3) the relevance for the core philosophical debate on individual-collective responsibility and agency.

The social physics case study tries to monitor and understand individual and crowd behaviours. We showed that ethical values such as responsibility, autonomy, privacy, and others, are in the models and the research. The sociophysics researchers use these values and their individual or collective characteristics implicitly in their models.

In future work, we will conduct interdisciplinary research on crowd management at train stations from sociophysics, psychology, and ethics of technology. Further support of sociophysics might help ethical research to get more empirical evidence about the relation between individual and collective behaviour to the ethics debate. As such, ethics research will be able to use the empirical information to formulate new insights on ethics in crowds. This will be particularly relevant in cases of COVID-19 regulations.

This brings us to the second lesson we believe can be drawn from the ethics of nudging of individuals and crowds. Now it may already seem clear that nudging all individuals to exert the same healthy behaviour is different from nudging some people to do different things, such as going left while others go right to disperse crowds. Therefore, some common and important issues pertaining to nudging are: (1) What exactly it means to 'nudge a crowd'? (2) How can the ethical rules governing individuals and crowds be separated?; (3) How can we nudge crowds from both an ethical and psychological point of view, while respecting values such as autonomy and privacy; (4) how can crowd properties (e.g. density, spread, flow) modify individual behaviour, for example, relevant to respecting COVID-19 rules.

Of course, the above discussions of the empirics and the ethics of nudging collectives bumps into the fundamental philosophical questions of individual-collective agency/responsibility, and whether collective agency, responsibility, deliberation, and values in general, exist or not. This is also relevant for obtaining an overall view of what (a) society is, which is particularly important if we zoom in on the application of COVID-19 measures.

Although this chapter is only exploratory, we postulate that further research in this direction might add to this fundamental debate. Further empirics and understanding of the interactions of crowds in a particular train-station situation might provide information on the fundamental interactions between individuals and collectives in society. It would help develop further guidelines for democratic decisions in crisis moments such as during the COVID-19 pandemic. This information will help specify how concrete measures should focus on individuals and collectives, and how to increase the effectiveness and the propriety of these measures.

Ultimately, our exploratory analysis above intends to emphasise the golden question for the post-COVID-19 debate, namely, *what is the right way to balance individual and collective values in the future.* This is, fundamentally, an ethico-philosophical debate, but which has wide-ranging effects on many other societal domains, such as health, economy, technology, and more.

In all, considering the above example advancements in crowd management techniques and tools, we argue that having a multidisciplinary and data-driven approach during ethico-philosophical argumentation and analysis can significantly enhance them. And since ethics and philosophy can be improved with the help of other data-driven and real-life studies, ethicists and philosophers can thus produce better argumentations regarding various public health and security policies during their design, enactment and implementation. Therefore, we argue that policymakers ought to engage more with ethicists and philosophers during the design of these policies, especially those that utilise multidisciplinary approaches.

Finally, taking all the above into account, we argue that there should be a widespread public debate on the balance between individual and collective values, general balancing principles in this regard, the assertion of primacy (conflict resolution) during and outside times of crises, the boundaries of governmental action in mandating measures, the acceptable use of technology, and the policy created thus.

This debate must include all relevant stakeholders (government, scientific institutions, the public, identified groups at an increased risk, medical practitioners, philosophers, ethicists, sociologists, psychologists, and others). It must result in policy deemed acceptable by all of the above to provide authority to policymakers and avoid the tension, polarisation, and perceived rise of (latent) authoritarianism recently observed among and by citizens in democratic societies.

Acknowledgements This work is part of the HTSM research programs "SRCrowd: Individual and collective agency in Socially Responsible nudging of Crowd", and "HTCrowd: a high-tech platform for human crowd flows monitoring, modeling and nudging" with project numbers 18754, 17962, and the VENI-AES research program "Understanding and controlling the flow of human crowds" with project number 16771, all financed by the Dutch Research Council (NWO).

Figures in this article are novel reproductions of the ones included in the original PLOS publication (Pouw et al., 2020).

References

Al Jazeera. (2021). Melbourne set to bring an end to world's longest lockdowns. *Coronavirus pandemic News | Al Jazeera*. https://www.aljazeera.com/news/2021/10/17/australias-melbourne-set-to-end-worlds-longest-lockdowns. Retrieved 2021-11-05.

Amer, F., Hammoud, S., Farran, B., Boncz, I., & Endrei, D. (2021). Assessment of countries' preparedness and lockdown effectiveness in fighting COVID-19. *Disaster Medicine and Public Health Preparedness, 15*, E15–E22. https://doi.org/10.1017/2Fdmp.2020.217

Asongu, S. A., Diop, S., & Nnanna, J. (2020). The geography of the effectiveness and consequences of Covid-19 measures: Global evidence. *Journal of Public Affairs*, e2483. https://doi.org/10.1002/pa.2483

BBC News. (2021). Coronavirus: Wuhan emerges from 76-day lockdown. *BBC News*. https://www.bbc.com/news/in-pictures-52215631. Retrieved 5 Nov 2021.

Beeckman, M., De Paepe, A., Van Alboom, M., Maes, S., Wauters, A., Baert, F., Kissi, A., Veirman, E., Van Ryckeghem, D. M. L., & Poppe, L. (2020). Adherence to the physical distancing measures during the COVID-19 pandemic: A HAPA-based perspective. *Applied Psychology: Health and Well-Being, 12*(4), 1224–1243. https://doi.org/10.1111/aphw.12242

Berg-Beckhoff, G., Guldager, J. D., Andersen, P. T., Stock, C., & Jervelund, S. S. (2021). What predicts adherence to governmental COVID-19 measures among Danish students? *International Journal of Environmental Research and Public Health, 18*(4), 1822. https://doi.org/10.3390/ijerph18041822

Borch, C. (2009). Body to body: On the political anatomy of crowds. *Sociological Theory, 27*(3), 271–290. https://doi.org/10.1111/2Fj.1467-9558.2009.01348.x

Capasso, M., & Umbrello, S. (2021). Responsible nudging for social good: new healthcare skills for AI-driven digital personal assistants. *Med Health Care and Philos 25*, 11–22. https://doi.org/10.1007/s11019-021-10062-z

Chiao, J. Y., & Blizinsky, K. D. (2010). Culture–gene coevolution of individualism–collectivism and the serotonin transporter gene. *Proceedings of the Royal Society B: Biological Sciences, 277*(1681), 529–537. https://doi.org/10.1098/rspb.2009.1650

Corlett, J. A. (2001). Collective moral responsibility. *Journal of Social Philosophy, 32*(4), 573–584. Cited in Smiley 2017 (see below). https://doi.org/10.1111/0047-2786.00115

Donnelly, J. (2013). Universal human rights in theory and practice. In Universal Human Rights in Theory and Practice. Cornell University Press.

Engelen, B., & Nys, T. (2020). Nudging and autonomy: Analyzing and alleviating the worries. *Review of Philosophy and Psychology, 11*(1), 137–156. https://doi.org/10.1007/s13164-019-00450-z

Feldman, S., & Stenner, K. (1997). Perceived threat and authoritarianism. *Political Psychology, 18*(4), 741–770. https://doi.org/10.1111/0162-895X.00077

Fogg, B. J. (2002). Persuasive technology: Using computers to change what we think and do. *Ubiquity, 2002*(December), 2.

Haidt, Jonathan. (2012). The righteous mind: Why good people are divided by politics and religion. Vintage.

Hausman, D. M., & Welch, B. (2010). Debate: To nudge or not to nudge. *Journal of Political Philosophy, 18*(1), 123–136. https://doi.org/10.1111/j.1467-9760.2009.00351.x

Hösle, V. (2004). *Morals and politics.* University of Notre Dame Press.

Hoxhaj, A., & Zhilla, F. (2021). The impact of covid-19 measures on the rule of law in the Western Balkans and the increase of authoritarianism. *European Journal of Comparative Law and Governance, 8*(4), 271–303.

Hozhabri, H., Sparascio, F. P., Sohrabi, H., Mousavifar, L., Roy, R., Scribano, D., De Luca, A., Ambrosi, C., & Sarshar, M. (2020). The global emergency of novel coronavirus (SARS-CoV-2): An update of the current status and forecasting. *International Journal of Environmental Research and Public Health, 17*(16), 5648. https://doi.org/10.3390/ijerph17165648

Ishay, M. (2020). *The history of human rights.* University of California Press.

Johanssen, J. (2016). The subject in the crowd: A critical discussion of Jodi Dean's "crowds and party". tripleC: Communication, capitalism & critique. *Open Access Journal for a Global Sustainable Information Society, 14*(2), 428–437.

John, P., Cotterill, S., Richardson, L., Moseley, A., Smith, G., Stoker, G., Wales, C., Liu, H., & Nomura, H. (2013). *Nudge, nudge, think, think: Experimenting with ways to change civic behaviour.* A&C Black.

Jones, D. R. (2015). Declining trust in congress: Effects of polarization and consequences for democracy. *The Forum, 13*(3), 75–394. https://doi.org/10.1515/for-2015-0027

Jungkunz, S. (2021). Political polarization during the COVID-19 pandemic. *Frontiers in Political Science, 3*, 6. https://doi.org/10.3389/fpos.2021.622512

Keiser, J. (2021). The Netherlands struggles to find political stability as polarisation increases. Global Risk Insights. *Global Risk Insights*, 22. https://globalriskinsights.com/2021/06/the-netherlands-struggles-to-find-political-stability-as-polarisation-increases/. Retrieved 2021-11-05.

Kimball & Josephs (2021). Businesses have until after the holidays to implement Biden Covid vaccine mandate. *CNBC.* CNBC, November 4, 2021. https://www.cnbc.com/2021/11/04/

biden-vaccine-mandate-businesses-have-until-after-christmas-to-comply.html. Retrieved 2021-11-05.

Klir, G. J. (2013). *Facets of systems science* (Vol. 7). Springer Science & Business Media, Springer. https://doi.org/10.1007/978-1-4615-1331-5

LibreTexts. (2020). *Durkheim and social integration.* https://socialsci.libretexts.org/@go/page/7896. Retrieved 2021-11-04.

Maine, H. S. (2007). *Ancient law its connection to the history of early society.* Project Gutenberg.

Maoz, Z., & Zeynep, S.-T. (2010). Political polarization and cabinet stability in multiparty systems: A social networks analysis of European parliaments, 1945-98. *British Journal of Political Science, 40*(4), 805–833. https://doi.org/10.1017/S0007123410000220

Novel, C. P. E. R. E. (2020). The epidemiological characteristics of an outbreak of 2019 novel coronavirus diseases (COVID-19) in China. *Zhonghua liuxingbingxue zazhi, 41*(2), 145. https://doi.org/10.3760/cma.j.issn.0254-6450.2020.02.003

Pouw, C. A. S., Toschi, F., van Schadewijk, F., & Corbetta, A. (2020). Monitoring physical distancing for crowd management: Real-time trajectory and group analysis. *PLoS One, 15*(10), e0240963. https://doi.org/10.1371/journal.pone.0240963

Schubert, C. (2017). Green nudges: Do they work? Are they ethical? *Ecological Economics, 132*(C), 329–342. https://econpapers.repec.org/scripts/redir.pf?u=https%3A%2F%2Fdoi.org%2F10.1016%252Fj.ecolecon.2016.11.009;h=repec:eee:ecolec:v:132:y:2017:i:c:p:329-342

Schwartz, S. H. (1992). Universals in the content and structure of values: Theoretical advances and empirical tests in 20 countries. *Advances in Experimental Social Psychology, 25*, 1–65. Academic Press.

Schwartz, S. H. (1994). Are there universal aspects in the structure and contents of human values? *Journal of Social Issues, 50*(4), 19–45. https://doi.org/10.1111/j.1540-4560.1994.tb01196.x

Schwartz, S. H. (2017). The refined theory of basic values. In *Values and behavior* (pp. 51–72). Cham.

Schweikard, D. P., & Schmid, H. B. (2021). Collective intentionality. In *The Stanford encyclopedia of philosophy.* (Fall 2021 Edition).

Sikali, K. (2020). The dangers of social distancing: How COVID-19 can reshape our social experience. *Journal of Community Psychology, 48*, 2435–2438. https://doi.org/10.1002/jcop.22430

Smiley, M. (2017). Collective responsibility. In *The Stanford encyclopedia of philosophy.* (summer 2017 edition).

Spahn, A. (2012). And lead us (not) into persuasion...? Persuasive technology and the ethics of communication. *Science and Engineering Ethics, 18*(4), 633–650. https://doi.org/10.1007/s11948-011-9278-y

Spahn, A. (2018). "The first generation to end poverty and the last to save the planet?"—Western individualism, human rights and the value of nature in the ethics of global sustainable development. *Sustainability, 10*(6), 1853. https://doi.org/10.3390/su10061853

Spahn, A. (2020). Digital objects, digital subjects and digital societies: Deontology in the age of digitalization. *Information, 11*(4), 228. https://doi.org/10.3390/info11040228

Thaler, R. H., & Sunstein, C. R. (2008). *Nudge: Improving decisions about health, wealth, and happiness.* Yale University Press.

The Guardian. (2021). Austria to put millions of unvaccinated people in Covid lockdown. *The Guardian.* Guardian News and media, (2021-11-12). https://www.theguardian.com/world/2021/nov/12/austria-province-to-place-millions-of-unvaccinated-people-in-covid-lockdown

Thomson, S., & Ip, E. C. (2020). COVID-19 emergency measures and the impending authoritarian pandemic. *Journal of Law and the Biosciences, 7*(1), lsaa064. https://doi.org/10.1093/jlb/lsaa064

Triandis, H. C. (1995). *Individualism & collectivism.* Westview Press.

United Nations. (1945). United Nations charter.

United Nations. (1949). The universal declaration of human rights.

Way, B. M., & Lieberman, M. D. (2010). Is there a genetic contribution to cultural differences? Collectivism, individualism and genetic markers of social sensitivity. *Social Cognitive and Affective Neuroscience, 5*(2–3), 203–211. https://doi.org/10.1093/scan/nsq059

Wright, L., Steptoe, A., & Fancourt, D. (2020). *What predicts adherence to COVID-19 government guidelines? Longitudinal analyses of 51,000 UK adults. MedRxiv.* https://doi.org/10.1101/2020.10.19.20215376

Open Access This chapter is licensed under the terms of the Creative Commons Attribution 4.0 International License (http://creativecommons.org/licenses/by/4.0/), which permits use, sharing, adaptation, distribution and reproduction in any medium or format, as long as you give appropriate credit to the original author(s) and the source, provide a link to the Creative Commons license and indicate if changes were made.

The images or other third party material in this chapter are included in the chapter's Creative Commons license, unless indicated otherwise in a credit line to the material. If material is not included in the chapter's Creative Commons license and your intended use is not permitted by statutory regulation or exceeds the permitted use, you will need to obtain permission directly from the copyright holder.

Chapter 13
Rhetorics of Resilience and Extended Crises: Reasoning in the Moral Situation of Our Post-Pandemic World

Samantha Copeland and Jose C. Cañizares-Gaztelu

13.1 Introduction

The normative discourses that have arisen around the COVID-19 global pandemic illustrate essential changes in our moral landscape. We argue in this chapter that these changes raise important moral challenges, but that some of these challenges can be at least partly addressed by critically assessing the role of resilience in pandemic discourse.

Since the 1970s in ecology (Holling, 1973), and increasingly in many other scientific disciplines and practical contexts (Brown, 2012; Doorn, 2015; Meerow & Stults, 2016), resilience has been proposed as a principle and approach for managing complex systems in a context of uncertainty. In many of these accounts, resilience is viewed as a descriptive concept that denotes some kind of response of complex systems to shocks and stresses (Brand & Jax, 2007; Elmqvist et al., 2019). However, tropes about resilience also became rather omnipresent during the pandemic, highlighting its complex, unexpected and unpredictable character, and communicating advice and instruction over what we can and should do in such an unusual situation. Because resilience has become an important concept for practical and moral reasoning in and about the pandemic, we look closely at the pitfalls and potential benefits of these normative uses of resilience in pandemic discourse.

We begin by addressing both the situation and the nature of the moral complexity elicited by the pandemic (Sect. 13.2). Next we introduce relevant conceptual aspects of resilience (Sect. 13.3) and illustrate some key and recurrent resilience tropes in the rhetoric around COVID-19 (Sect. 13.4). After taking up normativity theory to highlight and critically assess some problematic normative aspects of

S. Copeland (✉) · J. C. Cañizares-Gaztelu
Delft University of Technology, Delft, Netherlands
e-mail: s.m.copeland@tudelft.nl; j.c.canizaresgaztelu@tudelft.nl

© The Author(s) 2022   233
M. J. Dennis et al. (eds.), *Values for a Post-Pandemic Future*, Philosophy
of Engineering and Technology 40, https://doi.org/10.1007/978-3-031-08424-9_13

these tropes (Sect. 13.5), we suggest ways to overcome or at least address the conflicts and problems these tropes seem to raise (Sect. 13.6).

13.2   The Moral Situation Presented by the Pandemic

The pandemic presents us with a situation that is particularly riddled with moral complexity. For example, David Shaw (2021) characterizes this situation as one where we experience a lack of motivation to comply with imposed restrictions due to the problem of 'moral distance'. Shaw argues that the distance between us limits our ability to perceive or to address our moral duties to each other effectively, because we cannot properly assess the probable consequences of our actions. For example, asymptomatic individuals are unlikely to know they have the virus, and so their most rational and considerate assessments may still be incorrect: they may indirectly infect someone despite their best efforts to take precautions. This line of reasoning also clearly echoes the problem of 'moral luck',[1] wherein contingencies, rather than intentions or even causal relations, determine the moral evaluation of an action. For example, I may perform the same actions with the same intentions, such as going out to dinner and following the masking and distancing rules as required, and in one case dine without consequences, but in another case contribute to a cascade of infections that results in someone's death—depending, perhaps, on the weather and the way the wind was blowing that day.

The distance problem and the issue of moral luck illustrate the moral dilemmas that arise when we weigh our actions and choices against both their current and close and their distant and future implications. When taking a 'multi-scalar' perspective, apparently simple situations become complex and uncertain; when one cannot know for certain the results of one's actions, one cannot easily decide which actions will be the best or the 'right' thing to do. Here we want to argue that our situation as moral agents in the pandemic is still more complex, but also, not hopeless. Shaw himself proposes a strategy for increasing people's awareness of consequences that are probable even if unpredictable in this situation, but he focuses only on a limited subset of the problems involved in this multi-scalar moral situation, those that relate to our other-regarding decisions and actions. His solution, to provide more awareness of the probable and possible implications of our actions for others, is consequently insufficient to guide moral reasoning in this complex situation. We think that in the case of this pandemic, this picture needs broadening in at least three significant directions in order to enhance our understanding of the moral challenges at hand.

One relates to the nature of the pandemic *crisis* –a term that is both accurate and telling. The sudden and disruptive pandemic onset could be framed as a shock with

---

[1] Please note this is indeed a shallow review of two problems that philosophers have put considerable thought toward, but a deeper analysis is outside the scope of this chapter.

which we had to cope. But episodes such as the 3-month full lockdown in Spain are more suitably described as imposing ongoing stress upon individuals, households and local systems. In hindsight, rather than as a single stressor or shock, the pandemic as a whole is best characterized as having involved (and as still involving) bundles of stressors and shocks that compound and interact with each other across space and over time. One can learn from shocks and apply those lessons to similar shocks in the future. As illustrated by adaptive preferences (Elster, 1983; Teschl & Comim, 2005), one can also get used to stressors and learn to live with them. But how does one adapt to, and make decisions about, the radically uncertain (Hansson, 1996; Stirling, 2010) –where not only the likelihood of outcomes, but the possible outcomes themselves, and even the intervening factors in the situation, are unknown? In other words: we must accommodate both the many concrete and more or less tractable moral challenges that the pandemic is forcing upon us, and the general context of extended and evolving crisis that the pandemic itself represents.

A second issue is that our self-regarding decisions and actions (the prudential dimension of ethics) are also critical. Granted, we need to protect the health of those we know and, surely, we also have an obligation towards the welfare of those who are distant from us, in space or time. But this duty cannot be neatly separated from the duty to protect our own health by avoiding catching the virus –and then passing it along. Moreover, the pandemic also threw upon us many other economic and social problems with longer term and often more intangible repercussions: we struggle to cope with these problems because they can affect both us and others in a reciprocal fashion. Indeed, through countless media platforms, memes and news, in this pandemic we were bombarded with all kinds of tips for coping with the pandemic, the policy changes and the problems they entail, including the stresses of self-isolation, job loss and increasing duties at work and home (now overlapping for many of us), and even with the growing anxiety about impending global economic collapse. Thus, another key moral fact about the pandemic is that its repercussions are tangible and intangible, near and remote, and that they affect us and then others -and vice versa. These cross-scalar and iterative effects mean not only that we have self-regarding as well as other-regarding duties: in a sense, they mean that the distant other is also us.

Framing the situation in terms of moral distance alone also neglects the transformative potential of the pandemic. As we live through the pandemic, we struggle to cope with the problems we encounter. Yet, as the crisis persists and unfolds in new directions, we also try to create and seize opportunities to enact change that might enable us to respond better both to the pandemic and to similar crises in the future. Indeed, we have sometimes been asked to actively embrace the change forced upon the world for its transformative potential. For example, as Arundhati Roy argued early on, "[t]he pandemic is a portal" (Roy, 2020) –an opportunity to embrace radical change for climate mitigation and adaptation, now that the pandemic has demonstrated our capacity for accepting radical change, and because returning to "normal" is implausible at any rate.

Thus, the dilemmas with moral reasoning at various scales come in many forms in this situation. Can we prioritize ourselves against others, and should we? Is this travel policy a matter of health, of economic interest, of national identity, or of trust-building? Should it be different, and why? Can I afford sticking to conventions, to the law and scientific advice, or should I be bolder, and when? These dilemmas cannot be understated –in fact, they extend beyond the moral distance issue high-lighted by Shaw. Yet, in the next sections our position will be that lessons from resilience thinking can capture many of these moral dilemmas while also offering a guide for ethical deliberation and thought –in the context of the pandemic and beyond. To this end, we turn now to resilience research to briefly explain what this concept is about and some of the tensions and problems involved in its use.

13.3 The Nature of Resilience

Having its origins in the mechanical sciences, resilience is now used in multiple ways in many disciplines (Alexander, 2013), and is consequently both a complex and ambiguous concept overall (Brand & Jax, 2007; Strunz, 2012; Woods, 2015). Despite this variety of uses, however, classical accounts of resilience coincide in several ways. First, they generally present resilience as manifesting in conditions where uncertainty reigns: more particularly, as the ability to respond well and sur-vive through unpredictable or unforeseeable shocks or stressors (Holling, 1973, 1978; Norris et al., 2008). Second, resilience is applied at various scales: in psychol-ogy, for example, it is the individual propensity or demonstrated capacity to with-stand crises or shocks (Southwick et al., 2014); in ecology and related sciences, it denotes a similar capability, but of complex ecological systems, from the local (Hughes et al., 2005) to the global (Rockström et al., 2009). A third widely noted feature of resilience is the complicated interplay between conservation and change it denotes (Carpenter & Brock, 2008), since resilient individuals or systems are those that 'bounce back' from a crisis, but also adapt effectively to new circum-stances while retaining primary functions. For example, people are resilient insofar as they maintain (primarily physical and psychological) health despite encountering great adversity (Southwick et al., 2014), and/or if they adapt well to novel and unex-pected conditions (Norris et al., 2008); and ecological and other complex systems are resilient when they "absorb disturbance and reorganize while undergoing change so as to still retain essentially the same function, structure, identity, and feedbacks" (Walker et al., 2004). In general, then, in the context of a shock or stress, resilient systems respond by preserving their identity (or their critical features) while also leaving behind the non-essential, or adapting somehow to the new situation.

Although traditional approaches to resilience are still very influential (Elmqvist et al., 2019), resilience thinking has undergone an important evolution in the last two decades. One such development concerns a social turn in resilience thinking (Brand & Jax, 2007). At least since the mid-1990s, the ecological perspective on resilience has been proclaimed applicable to any complex system (Holling, 1996), which prompted efforts to understand and address the resilience of socio-ecological systems (Gunderson & Holling, 2002), engineering and socio-technical systems

(Hollnagel et al., 2006; Wardekker et al., 2010) and other systems of systems, including cities (Meerow et al., 2016). Today, resilience is widely applied in contexts such as urban planning or development studies, often by combining complex systems insights about natural systems, infrastructure, society and institutions into comprehensive strategies related to the management of risks.

This social turn has also raised the growing need to reconcile the system-of-systems perspective of resilience, coming from ecology, with the inclusion of people in this picture. Consider the example of cities. While cities can be framed as systems-of-systems or networks-of-networks, such perspectives might ignore individuals, and even communities and their identity or culture (Meerow et al., 2016). Questions such as *resilience of what to what*, or *resilience for whom*? (ibid) become, then, quite pressing, especially for specifying the so-called critical features that stand for the "identity" of the system of interest (viz. above). For example, when Hurricane Katrina devastated the cultural core of New Orleans in 2004, questions were raised about how to build more resilience into the recovering city: was it more important to maintain the structures of the city exactly as they were, preserving neighbourhoods that were culturally significant, and to ensure that the people could come back to the neighbourhoods they lived in before the disaster? Or is the overall resilience of the city structure more important, so that some vulnerable neighborhoods might have to be sacrificed to rebuild better elsewhere, preserving the city's population but trading away its historical ties? (Kates et al., 2006).

Another important development has to do with the kind of disturbance that resilient systems are supposed to be resilient to. Resilience had been initially applied to *specific* kinds of shocks (sudden and disruptive events) and stresses (long-onset and persistent disturbances upon normal or typical performance). However, following the social turn, resilience began to be interpreted as a more general capacity to withstand various kinds of uncertain stresses and shocks, or combinations of them, at various scales and over an indefinite period –what is known as *general* resilience (Carpenter et al., 2012). General resilience has increasingly attracted attention in contexts such as urban adaptation to climate change or risk management (Cañizares et al., 2021), where the concern is not primarily with single stressors or shocks, but rather with bundles of stressors that appear and disappear or become latent, spanning from the individual to (immediately, through spillovers and cascading effects) the global. Consequently, it is nowadays common to find multi-scalar and general approaches to the resilience of, for example, communities, cities or economies (Norris et al., 2008; Rockefeller Foundation and Arup, 2016).

Increasingly, too, approaches to resilience have become more forward-looking, sometimes captured as 'bouncing forward', or transformative, rather than bouncing back (Bahadur & Tanner, 2014). While classic accounts of resilience had already noted that resilience is not mere resistance (Carpenter et al., 2001), recent accounts insist more on the dynamic nature of resilience. It is now accepted that efforts to develop resilience must account for the change that will inevitably occur when responding to a crisis, and moreover, that it is neither possible nor always desirable to return to the previous status quo (Copeland et al., 2020). The features that caused a collapse in a flood protection system, for instance, cannot simply be repaired since the original system was demonstrably not resilient. Thus, the concept of resilience denotes two complementary but potentially competing challenges in dealing with "disturbances": the need to prevent collapse by preserving critical functions or features, and the need to change, transform or be adaptable in order to allow for more effective responses to future disturbances (Meerow et al., 2016). Efforts to build resilience can represent conservative measures toward preservation as well as transformative measures to enact necessary changes.

A further important development concerns the normative use of resilience. Prominent accounts of ecological and socio-ecological resilience had tended to portray resilience as a descriptive concept –a property of complex systems in general, which can be good or bad, desirable or not: see, e.g. the above quoted definition by Walker et al. (2004); also (Anderies et al., 2013; Elmqvist et al., 2019). Recently, however, this characterization of resilience has been criticized as incoherent, since, in most if not all its applications, resilience is used as a goal or principle for framing and guiding risk management strategies (Cañizares et al., 2021). This is especially the case in social applications of resilience, which necessarily involve explicitly normative decisions and, moreover, tend to frame resilience as a positive feature or ability (Olsson et al., 2015; Meerow et al., 2016; Thorén & Olsson, 2017).

The next sections return to these topics, especially to questions and concerns about the normativity of resilience. Now we present some tropes of resilience that became quite common during the pandemic. These tropes illustrate the diverse uses to which resilience can be put, as well as some of the tensions that typically underlie usage of this term.

13.4   Resilience Tropes in the Pandemic

Since the pandemic was announced in 2020, we have seen several common tropes arise in media discourse and in the rationales for the policy approaches taken by institutions. Resilience has occupied a prominent place within these discourses. As individuals who find our behaviour mandated by such policies, we have been called upon to help and to 'build resilience' in at least three different ways. First, on the personal level, we are guided toward resources that will help us resist the virus and cope with the disruptions that policies such as self-isolation bring to our lives. Second, the social resilience of our communities, cultures and countries, is affected by our individual behavior, which is in turn mandated to enable group-level resilience. Third, on a higher level, the resilience of the human species has been part of debates about policy, and even more so the resilience of our institutions and society as a whole are threatened by the pandemic; certain ways of behaving, we are told, will help us return to 'normal' more quickly, where 'normal' might mean the freedom to travel, living our social lives, and even returning to the economic stability that many people had and lost with the pandemic.

Individual or personal resilience has been framed in the pandemic discourse both in terms of biological and psychological well-being. In some cases, it rather straightforwardly refers to physical resilience to the COVID-19 virus and its effects; are individuals healthy and strong enough to suffer from and yet survive both the virus and its knock-on effects? Indeed, some groups are seen as naturally more or less resilient to the effects of COVID-19 and the pandemic countermeasures than others. For example, consider the impact of the pandemic on children who have had to miss education and important social development time with their peers as a result of school and playground closures for extended periods. The phrase 'kids are resilient'[2] has been used to suggest that children's inherent flexibility and ability to adapt will enable them to cope well enough with the changes to their lives required by pandemic restrictions. This trope is also present in various forms of advice given to employees or citizens by their employers or national institutions to be resilient in the face of the challenges brought by the pandemic and related policies. The Mental Health Commission of Canada Working Minds blog, for instance, reminds workers in its 'Self-care Resilience Guide' that, "this is a good time to remember...that you have resiliency skills and you can cope".[3] Likewise, the Centre for Disease Control in the U.S. offers individuals a number of "tips to build resilience and manage job stress," such as "Remind yourself that everyone is in an unusual situation with limited resources."[4]

Even a fairly straightforward reference to individual bodily health, however, also has a social and cultural context. Some groups have demonstrated greater physical resilience in response to the virus, such as those who already have 'killer T cells' remaining from a previous, less dangerous infection (Joy, 2021, in reference to Mallajosyula et al., 2021). Resilience to the virus, and also resilience to the impact of the pandemic as a whole, however, has more often been the consequence of the socio-economic context than of purely biological traits of those groups (Strang et al., 2020; Qureshi, 2021). Thus, the conception of personal resilience here entails the ability to cope well with the broader effects of the pandemic, such as stress, isolation and its economic impact, social determinants of health that in turn affect biological resilience to disease as well. What generally unites these approaches is that they characterize resilience as an available resource that each one of us should be able to draw on.

This reference to the social and cultural context takes us to a second trope, which is rather focused on social resilience, i.e. the resilience of groups or communities. As members of these communities, we are asked to behave in ways that protect the

---

[2] For example, as a teacher in the U.S. said in relation to the topic of schools reopening: "It will be a community, and it's not ideal, but to keep people safe, it is what it is...Kids are resilient, and kids are adaptable." Retrieved August 2021 from https://www.alligator.org/article/2020/07/kids-are-resilient-students-and-teachers-respond-to-acps-reopening-plan

[3] Retrieved September 2021: Staying Resilience During the COVID-19 Pandemic, Working Minds blog: https://theworkingmind.ca/COVID19-tim; Webpage for the Working Mind COVID-19 Self-care and Resilience Guide: https://theworkingmind.ca/blog/working-mind-COVID-19-self-care-resilience-guide/

[4] "Employees: How to cope with job stress and build resilience during the COVID-19 pandemic" Updated Dec.23, 2020, Retrieved August 8, 2021. https://www.cdc.gov/coronavirus/2019-ncov/community/mental-health-non-healthcare.html

more vulnerable, for example: public mask-wearing as a community-wide mandate ensures that otherwise vulnerable individuals are better protected when they need to travel. Vaccinating oneself contributes to the overall resilience of the group, as well: at the time of writing, the most recent 'Bloomberg COVID Resilience Ranking', granted Ireland top spot as 'best place to be during the pandemic' because of its high rates of vaccination and policies promoting more social freedoms to the already immunized. The collective action required for pandemic policies to work thus falls under this resilience trope. For instance, again from the 'tips to build resilience', the CDC in the United States recommends: "Remind yourself that each of us has a crucial role in fighting this pandemic." Consequently, we are asked both to build our individual resilience by using the resources available to us, and also to contribute through our individual behavior to building resilience at the community level.

At a more abstract level and with pronounced future-oriented intent, tropes of resilience also call on us to behave or implement policies in ways that would contribute to the resilience of human society, our institutions, and even of certain global social-economic values. One point of debate about national policies has centered around whether certain approaches in pandemic response were aimed at the goal of so-called 'herd immunity'—while this wasn't a resilience-based trope *per se*, it does reflect the belief that nations and even the species could be more or less resilient in the future to COVID-19, depending on how we build immunity into the population now. The idea of herd immunity has a straightforward and unproblematic epidemiological rationale insofar as it relates to high vaccination rates –when most of the population is vaccinated, the herd as a whole gets immune. What made it a (problematic) novelty in the context of COVID-19 was that herd immunity approaches were advocated at a time when vaccines against this virus were not yet available. This particular interpretation of 'herd immunity' suggested that it might be necessary to allow for some sacrifice of the vulnerable now, in order to gain resilience to the virus at the population level in the future, and it was strongly opposed on both epidemiological and moral grounds (Napier, 2020). Scott Atlas was heavily criticized, for example, for suggesting in his role as advisor of the Trump administration that letting "a lot of people get infected" was an effective strategy for building immunity in the population overall. UK prime minister Boris Johnson was similarly lambasted early on in the pandemic by the president of the British Society for Immunology, for proposing herd immunity as a national strategy.

More direct references to resilience are found in countless articles on the resilience of supply chains, healthcare systems, businesses and other institutions that have been disrupted by the pandemic and, apparently, exposed as insufficiently resilient. Since the coronavirus took to the international stage in 2020, for example, dozens of articles have been published on the topic of the resilience of healthcare systems to pandemics –see e.g. Chaturvedi and Siwan (2020); Wang et al. (2020); Sundararaman et al. (2021); Saulnier et al. (2021). We also mentioned the Bloomberg COVID Resilience Ranking, a regularly revised evaluation of national strategies for dealing with the pandemic, which relies on indicators for healthcare quality, vaccination levels in the population, mortality rates and progress in terms of reopening borders to travel and trade, to assess "where the virus is being handled the most effectively with the least social and economic upheaval."[5] National strategies such as recently announced in the UK are also explicitly turning to resilience as a leading value. Common in the rhetoric of this last trope, therefore, is a focus on system or population level resilience, with a future orientation to using the pandemic as a corrective lesson or for preparing better to avoid similar trouble in the future.

We think that a critical view of resilience could have two normative functions in the pandemic and in similar situations: characterizing the salient moral challenges in this context, and offering some moral guidance for addressing them. To show how, we must first unpack and critically discuss the normative character of these tropes.

13.5   The Normativity of Resilience

As was noted in Sect. 13.2, resilience research features some disputes about whether this term is descriptive or normative. Those who view resilience as a descriptive term often refer to the fact that resilience can denote both positive and negative, moral and immoral, phenomena – there are resilient ecosystems, but also resilient tyrannies (Anderies et al., 2013). While it is unclear that this argument suffices for situating resilience as descriptive (Cañizares et al., 2021), the argument is nonetheless irrelevant in the pandemic context – the tropes of resilience reviewed above present it as a positive feature, and so, as an evaluative term. Moreover, these understandings of resilience are also generally used for implicitly or explicitly making prescriptions.

To explain, evaluative terms are those commonly used for ascribing a positive or negative valence or value to what they describe (Tappolet, 2013). For example, when we say something is beautiful or ugly, we judge it in an aesthetic sense to be good or bad, as having value or not. Virtues and vices are familiar categories of evaluative terms: when we say that someone has the virtue of generosity, we appraise her positively; someone with the vice of meanness is being appraised negatively. Generosity comes from good motives and reasons and leads to good outcomes— without these aspects, giving away one's money would be frivolous, or if it led to a bad end, irresponsible, rather than indicating the virtuous generosity of the one giving it away. It is typical for evaluative terms to be used to give reasons in favour or against something; it is typically the case that if we assess some action or event as good, we have reasons for doing so and would like it to happen or to be that way. Likewise, assessing something as bad goes hand in hand with its being undesirable. Evaluative language can be used thus to 'straddle the divide' between is and ought when an evaluation (an 'is') becomes the basis for a prescription (an 'ought').

---

[5] Retrieved in October 2021, but at the time of writing, the site is still being updated regularly here: https://www.bloomberg.com/graphics/COVID-resilience-ranking/

Note that these normative aspects are not always as transparent as they should be. This is most clearly exemplified by the first two tropes explored above, personal resilience and social resilience. In its more medical or biological interpretation, the trope of personal resilience denotes that someone has returned to full health, or that their body and mind have the capacity for responding effectively to viral invasion and the pandemic. More broadly speaking, however, this trope also refers to the resources available to us to care for our mental health and cope with the stresses of lockdown and other changes. The second trope is, as we saw, slightly different: it refers to our ability to harness our individual resilience and put it in service of our community.

Insofar as these tropes refer primarily to the observable signs of resilience, to a naturally occurring property of individuals or groups, or to how possessing certain features tends to result in a resilient outcome, here we might seem to be dealing with a descriptive category. Yet, note that these resources and our ability to harness them are both viewed as positive, insofar as they allow us (or our relatives and communities) to survive, maintain integrity and thrive. Consequently, these tropes are clearly evaluative. At the same time, they are also often used prescriptively, as when we are asked to draw on these resources in order to fight the pandemic, or when we say that 'kids are resilient' to advance or justify policies, for instance that prevent them from playing at playgrounds, or advise on their return to school, in favour of allowing other sectors of the economy to open.[6]

Precisely due to its normative implications, in contexts outside the pandemic, this trope of personal resilience has encountered considerable resistance. One common argument against it is that it allows for moral passivity toward the difficulties certain groups endure. For instance, a paper sign quoting Tracy L. Washington, stapled to a lamppost by the Louisiana Justice Institute in New Orleans after Hurricane Katrina, declares: "Don't call me resilient, Because every time you say, 'Oh, they're resilient,' that means you can do something else to me. I am not resilient." This trope is also critically portrayed as an intent to escape collective or institutional responsibility for improving social conditions by shifting the responsibility for ensuring resilience away from governing bodies and onto the shoulders of individuals. Psychologist and resilience researcher Michael Ungar (May 2019) put it bluntly in a short essay in the Canadian newspaper, the Globe and Mail: "The notion that your resilience is your problem alone is ideology, not science." Making people responsible for their own resilience is misdirected when their lack of resilience results mostly or even in part from social conditions that are best addressed at higher levels. It is also morally problematic when individuals do not really have the capability of being (more) resilient—that is, when the 'ought to be resilient' is not accompanied by the necessary 'is'. Those points of critique apply even more to the second trope, since social resilience is in many ways a matter of multi-level responsibility, from neighborhood to multilateral international governance, rather than just

---

[6] https://www.macleans.ca/society/health/the-pandemic-is-breaking-parents/

one of personal responsibility. Joseph (2013) has summarized these concerns most sharply by casting resilience as sheer neoliberal jargon.

The normative character of resilience is perhaps more explicit in instances of the third trope, where resilience is viewed as an ideal that the system of interest ought to attain, or progress toward. For instance, a resilient city could be one that is able to maintain what have been deemed its essential features, or one that is capable of improving or growing (progressing) in the face of disturbance. These understandings of urban resilience are quite different, but both are normative. In the former, resilience is about the conservation of something that is assumed to be good. In the latter, it is about transforming in order to improve. Such claims present resilience as a social or political value, that is, a desirable outcome or goal that institutions and systems like cities ought to strive for. Alternatively, resilience is often presented as a virtue: a desirable property of cultures, social organizations or ways of governance. One clear example of this use is the Bloomberg Ranking, whereby countries are deemed better or worse "places to be" during different phases of the pandemic, according to their criteria for handling the virus "most effectively with the least…disruption." Similarly, organisational theorists have written much about what makes for 'resilient leadership' through the pandemic, which illustrates the interpretation of resilience as an ideal or virtue of good governance, organization or business performance (Giustiniano et al., 2020).

These straightforward applications of systems views of resilience to social contexts have also been met with substantial criticism elsewhere, in light of their normative implications. In the development and climate adaptation literature, for instance, it has been claimed that the "apolitical systems perspective" conceals the normative character of resilience (Bahadur & Tanner, 2014). This is held to be morally problematic, since it contributes to depoliticizing resilience-based measures and to promoting a technocratic and managerial mindset that elides possible tradeoffs entailed by their application (ibid). Relatedly, some critics note that these perspectives tend to focus on systems properly speaking, such as e.g. in infrastructure or governance systems, while neglecting questions of power, rights of access to goods, and the differential impact of resilience-based measures and policy (Ziervogel et al., 2017). That has led some to question and even reject the idea that we should apply resilience to social contexts, since a return to even an undesirable status quo could be thereby sanctioned as a success (Béné et al., 2012). Scholars in this tradition therefore stress the need to be more explicit about the normative aspects of these system perspectives, especially by engaging with the aforementioned question of resilience for *whom*: who are the beneficiaries of resilience building, and who will be negatively affected by it (Meerow et al., 2016).

Recently, considerations of this sort have in fact prompted a wave of ethical and justice work in resilience research (Bulkeley et al., 2014; Shi et al., 2016; Fitzgibbons & Mitchell, 2019). In line with this work, we argue that making the normativity of the resilience we value explicit—as a set of evaluations that can lead to conflicting prescriptions for action—allows at least for deliberation about the priorities thereby set. Now we will look at how these uses of resilience can both confuse and have the potential to clarify the moral situation at hand in this pandemic.

13.6   Reasoning About and Towards Resilience
in the Pandemic Moral Situation

The resilience tropes around the pandemic, we suggest in this section, reflect the fact that we must engage multiple 'scales' when reasoning about our behaviour. As resilience is applied to individuals, groups and systems, these tropes advise us to consider factors at diverse levels and concerning different temporal ranges when deciding how we should behave. On the one hand, we must not only consider self-regarding, prudential reasons for our behaviour, but also other-regarding moral reasons at the same time. On the other hand, we are also consistently faced with the dual notions of transformation and preservation. That is, at the same time as we are dealing with current shocks and stressors, we are considering how we ought to improve ourselves and our systems so that this doesn't happen again (or continue to happen) in the future. These different scales of size and temporality make practical and moral reasoning particularly complex in contexts where iterative shocks and stressors are experienced with an uncertain end and where uncertainty about probable outcomes prevails.

To begin at the systems level, the concerns raised in the last section are somewhat condensed in the case of the idea of population resilience garnered via 'herd immunity'. As we noted, this was the idea that the survival of the majority of the population could be ultimately achieved by ensuring general immunity to the virus. Like the trope of personal resilience, this theme engages with the idea of survival as a naturally occurring property or ideal, and consequently seems like a simply descriptive category, but it is not. The survival of the numerical majority of a population is, of course, something that we would commonly evaluate as positive or desirable. In addition, the herd immunity approach implicitly prescribes some actions and inactions that are assumed to bring about immunity, such as increasing vaccination rates (the classical epidemiological approach) or limiting the social and institutional interference in people's normal lives (Sweden's and Boris Johnson's infamous approach). That is, resilience as herd immunity is not a naturally occurring or emergent ideal, but a reflection of the priorities we set and of our efforts toward ensuring them.

Yet, the way in which these priorities are set make the goal of herd immunity susceptible to the same objections raised against systems perspectives of resilience more generally. This could be expected, since herd immunity is, in general, a high-level social goal, and moreover one that does not always correlate with positive individual outcomes. Particularly, as Atlas and others (polemically) interpreted herd immunity in the pandemic onset, this idea means that the survival of the majority could be more likely if citizens were to go about their daily lives. By thus promoting herd immunity as a policy goal, then, not only the risks imposed on individuals are minimized, but, indeed, risk-taking social behavior is explicitly promoted among the population. In other words, the rhetorics of herd immunity imply, and at the same time they conceal, a clear conflict between system goals and personal and community values. Moreover, as we saw, survival is not a matter of simple bodily tolerance to the virus, but is, instead, heavily influenced by socio-economic circumstances. Thus, this case is one where questions over the potential tradeoffs between systems and individual perspectives on governance and policy are particularly critical, and yet in the name of resilience they may be elided, resulting in an intolerable neglect of precisely those who are most vulnerable to the virus and the pandemic in general. The solution here is to avoid using these rhetorics (about resilience or herd immunity) with a descriptive intent, and, instead, to explicitly unpack the normative impact these ideas have when we set them as goals.

At the personal and social levels, there is a range of factors relevant to our moral reasoning about behaviours like self-isolation and its consequences, such as not travelling to see family or moving one's social life online; we ought to consider the impact of those behaviours not only on ourselves and those to whom one usually is morally indebted, but also to the broader public and even the world. As we saw above, resilience is not only a positive characteristic for people to have during the pandemic—individuals are *called upon* to use the tools at their disposal to be more resilient—it is prescribed as a duty, while also describing a characteristic. Yet, while we may assess individuals as resilient or not, if they are not really capable of being more resilient on their own, nor should they thus be fully responsible for that resilience. While each of us is coping with reduced resources and difficulties during the pandemic, these hardships are not evenly distributed nor can they all be coped with well, without sufficient support. Contemporary approaches rather regard personal resilience as a reflection of capabilities and context rather than as an innate resource we can each call up when called upon (Norris et al., 2008). In this way, personal resilience is bound up with the resilience of social groups and systems level institutions: they interact.

Unpacking the normativity of resilience in rhetorical tropes such as the ones we have examined here is a first step toward understanding the moral complexity of the situation we are in. In the literature, as we say above, it has been suggested that unpacking the content of 'resilience' requires asking further questions, namely, resilience to what, of what, and resilience for whom. Asking these questions allows us to deliberate about the evaluative and prescriptive elements of resilience when it is applied as a trope to guide or advise us on how to conceptualise and to cope with the pandemic. Further, they provide a means to address the complexity of the decisions and choices that need to be made about what actions ought to be taken. We show here how the use of resilience in the pandemic rhetoric reveals the different levels on which we must reason about our behavior; as a value or goal, resilience represents the particular moral situation in which we must reason during a pandemic. Consequently, by making its normativity explicit, resilience becomes not only a way to evaluate our behavior, but a frame within which we can deliberate about what we should preserve, about ourselves and about the systems we can influence, and what we should change.

Consider further our early example, of deciding whether to go out to dinner, which requires assessing more than one risk, including risks that one cannot predict. Individuals evaluate their role as potential viral vectors in the pandemic and their social roles, the roles they play as workers, family members, and citizens. Individuals must consider the changing grounds of policy, science, medicine and resource availability, as well as their own needs and the needs of others who depend upon them. People need to consider factors on 'multiple scales' at the same time, temporally and in terms of systems: we need to consider our future while protecting ourselves in the present; we are both individuals and more or less essential parts of a larger ecological, social, economic and technical system. Depending on which scale we might focus on, different decisions will appear morally correct, and it is not unusual for alternatives to conflict. In all cases, the individual remains uncertain about the actual effects of their actions because COVID-19 transmission and its effects can be unpredictable. While this kind of complexity in moral reasoning is not novel, understanding why and how we value resilience in the context of an extended crisis, we suggest, shows us how complex systems can offer more than one and sometimes conflicting options for right action, as well as how we might go about deciding between them.

This moral complexity is illustrated when different answers to 'resilience to what' are considered, as they lead to differing responses to 'resilience for whom', for instance. To follow lockdown restrictions, for example, resilience to the aggregative effects of self-isolation will be required. This kind of policy, in fact, more or less takes the resilience of individuals to the impact of self-isolation to be a necessary requirement, in order to build a resilient society that also includes vulnerable people (whose risks are in turn intentionally reduced by that policy). This is in sharp contrast to policies like the so-called 'herd immunity' approach described above, which proposes instead to ignore the vulnerable in favour of building (a different kind of) resilience for the majority. Examining these policies by differentiating between the normative implications of 'resilience' used to promote or explain them, does the work of highlighting the alternatives we have for setting priorities, and their implications for the people involved.

Further, it is necessary to answer the questions, resilience to what, of what, and resilience for whom, to deliberate about what elements in the current system—or features of our current selves—we ought to keep and which ones we should change, given the opportunity to improve. By taking up an explicitly evaluative approach, the answers to these questions will help elucidate the nature of the evaluations we are making and the consequent prescriptions implied. Trade-offs are generally required for resilience, and depending on what they must be resilient to, the what and for whom resilience is a goal will differ. Like the survivors of a pandemic who now have 'herd immunity', the city that is deemed resilient in the aftermath of a crisis reflects choices made before and during that crisis about who and what constitutes that city's identity. In either case, it is possible and essential to deliberate explicitly about the evaluations we are making and their normative weight in terms of the prescriptions they imply.

13.7   Conclusion

Resilience has been applied as a concept and a value in the pandemic and elsewhere. Here we have shown that resilience thinking indeed has much to offer by way of highlighting morally relevant aspects of the pandemic and offering some guidance to moral reasoning in this context. However, as we saw, resilience is not without problems. Here we showed that resilience is a normative concept that is applied at various scales to denote conservation as well as transformation. Due to these features, resilience raises various concerns, for example: what are the things or properties to be conserved and which should be transformed? Who are the beneficiaries and the losers of resilience building? Can high-level systems such as nations be resilient if their citizens are not, and conversely, can we afford to neglect the context and support needed to build personal resilience? As we showed in our analysis of resilience tropes, failure to address these questions may mean missing opportunities for transformation, creating or reproducing tradeoffs between individual resilience and resilience at higher levels, and ultimately losing the potential of this concept for guiding critical and sensitive reflection over the great social challenges that lie ahead.

References

Alexander, D. E. (2013). Resilience and disaster risk reduction: An etymological journey. *Natural Hazards and Earth System Sciences, 13*(11), 2707–2716. https://doi.org/10.5194/nhess-13-2707-2013

Anderies, J. M., Folke, C., Walker, B., & Ostrom, E. (2013). Aligning key concepts for global change policy: Robustness, resilience, and sustainability. *Ecology and Society, 18*(8). https://doi.org/10.5751/ES-05178-180208

Bahadur, A., & Tanner, T. (2014). Transformational resilience thinking: Putting people, power and politics at the heart of urban climate resilience. *Environment and Urbanization, 26*(1), 200–214. https://doi.org/10.1177/0956247814522154

Béné, C., Wood, R. G., Newsham, A., & Davies, M. (2012). Resilience: New utopia or new tyranny? Reflection about the potentials and limits of the concept of resilience in relation to vulnerability reduction Programmes. *IDS Working Papers, 2012*(405), 1–61. https://doi.org/10.1111/j.2040-0209.2012.00405.x

Brand, F. S., & Jax, K. (2007). Focusing the meaning(s) of resilience: Resilience as a descriptive concept and a boundary object. *Ecology and Society, 12*(1). https://doi.org/10.5751/es-02029-120123

Brown, K. (2012). Policy discourses of resilience. In M. Pelling, D. Manuel-Navarrete, & M. Redclift (Eds.), *climate change and the crisis of capitalism* (pp. 37–50). Routledge.

Bulkeley, H., Edwards, G. A. S., & Fuller, S. (2014). Contesting climate justice in the city: Examining politics and practice in urban climate change experiments. *Global Environmental Change, 25*, 31–40. https://doi.org/10.1016/j.gloenvcha.2014.01.009

Cañizares, J. C., Copeland, S. M., & Doorn, N. (2021). Making sense of resilience. *Sustainability, 13*(15), 8538.

Carpenter, S., Walker, B., Anderies, J. M., & Abel, N. (2001). From Metaphor to Measurement: Resilience of What to What?. *Ecosystems 4*(8), 765–781. https://doi.org/10.1007/s10021-001-0045-9

Carpenter, S. R., & Brock, W. A. (2008). Adaptive capacity and traps. *Ecology and Society, 13*(40).

Carpenter, S. R., Arrow, K. J., Barrett, S., Biggs, R., Brock, W. A., Crépin, A.-S., ... Zeeuw, A. D. (2012). General resilience to cope with extreme events. *Sustainability, 4*(12), 3248–3259.

Chaturvedi, M., & Siwan, R. M. (2020). Resilience of healthcare system to outbreaks. In *Integrated risk of pandemic: COVID-19 impacts, resilience and recommendations* (pp. 397–412). Springer Singapore.

Copeland, S., Comes, T., Bach, S., Nagenborg, M., Schulte, Y., & Doorn, N. (2020). Measuring social resilience: Trade-offs, challenges and opportunities for indicator models in transforming societies. *International Journal of Disaster Risk Reduction, 51*. https://doi.org/10.1016/j.ijdrr.2020.101799

Doorn, N. (2015). Resilience indicators: Opportunities for including distributive justice concerns in disaster management. *Journal of Risk Research, 20*(6), 711–731. https://doi.org/10.1080/13669877.2015.1100662

Elmqvist, T., Andersson, E., Frantzeskaki, N., McPhearson, T., Gaffney, O., Takeuchi, K., & Folke, C. (2019). Sustainability and resilience for transformation in the urban century. *Nature Sustainability, 2*. https://doi.org/10.1038/s41893-019-0250-1

Elster, J. (1983). *Sour grapes: Studies in the subversion of rationality*: Editions De La Maison des Sciences De L'Homme.

Fitzgibbons, J., & Mitchell, C. L. (2019). Just urban futures? Exploring equity in "100 resilient cities". *World Development, 122*, 648–659. https://doi.org/10.1016/j.worlddev.2019.06.021

Giustiniano, L., Cunha, M. P., Simpson, A. V., Rego, A., & Clegg, S. (2020). Resilient leadership as paradox work: Notes from COVID-19. *Management and Organization Review, 16*(5), 971–975.

Gunderson, L., & Holling, C. S. (2002). *Panarchy: Understanding transformations in human and natural systems*. Island Press.

Hansson, S. O. (1996). Decision making under great uncertainty. *Philosophy of the Social Sciences, 26*(3), 369–386. https://doi.org/10.1177/004839319602600304

Holling, C. S. (1973). Resilience and stability of ecological systems. *Annual Review of Ecology and Systematics, 4*(1), 1–23. https://doi.org/10.1146/annurev.es.04.110173.000245

Holling, C. S. (1978). *Adaptive environmental assessment and management*. John Wiley.

Holling, C. S. (1996). Engineering resilience versus ecological resilience. In P. E. Schulze (Ed.), *Engineering within ecological constraints* (pp. 31–43). National Academy Press.

Hollnagel, E., Woods, D. D., & Leveson, N. (2006). *Resilience engineering: Concepts and precepts*. Ashgate Publishing Limited.

Hughes, T. P., Bellwood, D. R., Folke, C., Steneck, R. S., & Wilson, J. (2005). New paradigms for supporting the resilience of marine ecosystems. *Trends in Ecology and Evolution, 20*(7), 380–386. https://doi.org/10.1016/j.tree.2005.03.022

Joseph, J. (2013). Resilience as embedded neoliberalism: A governmentality approach. *Resilience, 1*(1), 38–52. https://doi.org/10.1080/21693293.2013.765741

Joy, H. (2021). Here's why some people are more resilient to Covid-19. July 3. https://www.medindia.net/news/heres-why-some-people-are-more-resilient-to-covid-19-202076-1.htm

Kates, R. W., Colten, C. E., Laska, S., & Leatherman, S. P. (2006). Reconstruction of New Orleans after hurricane Katrina: A research perspective. *Proceedings of the National Academy of Sciences of the United States of America, 103*(40), 14653–14660. https://doi.org/10.1073/pnas.0605726103

Mallajosyula, V., Ganjavi., Chakraborty, S., McSween, A. M., Pavlovitch-Bedzyk, A. J., Wilhelmy, J., Nau, A., Manohar, M., Nadeau, K. C., & Davis, M. M. (2021). CD8+ T cells specific for conserved coronavirus epitopes correlate with milder disease in patients with COVID-19. *Science Immunology, 6*(61). https://doi.org/10.1126/sciimmunol.abg5669

Meerow, S., & Stults, M. (2016). Comparing conceptualizations of urban climate resilience in theory and practice. *Sustainability, 8*(7), 701.

Meerow, S., Newell, J. P., & Stults, M. (2016). Defining urban resilience: A review. *Landscape and Urban Planning, 147*, 38–49. https://doi.org/10.1016/j.landurbplan.2015.11.011

Napier, A. D. (2020). I heard it through the grapevine: On herd immunity and why it is important. *Anthropology Today, 36*(3), 3–7. https://doi.org/10.1111/1467-8322.12572

Norris, F., Stevens, S., Pfefferbaum, B., Wyche, K., & Pfefferbaum, R. (2008). Community resilience as a metaphor, theory, set of capacities, and strategy for disaster readiness. *American Journal of Community Psychology, 41*, 127–150. https://doi.org/10.1007/s10464-007-9156-6

Olsson, L., Jerneck, A., Thoren, H., Persson, J., & O'Byrne, D. (2015). Why resilience is unappealing to social science: Theoretical and empirical investigations of the scientific use of resilience. *Science Advances, 1*(4), e1400217. https://doi.org/10.1126/sciadv.1400217

Qureshi, S. (2021). Pandemics within the pandemic: Confronting socio-economic inequities in a datafied world. *Information Technology for Development, 27*(2), 151–170. https://doi.org/1 0.1080/02681102.2021.1911020

Rockefeller Foundation, & Arup. (2016). *City resilience index: Understanding and measuring City resilience*. Retrieved from Available online: https://assets.rockefellerfoundation.org/app/uploads/20171206110244/170223_CRIBrochure.pdf

Rockström, J., Steffen, W., Noone, K., Persson, Å., Chapin, F. S., III, Lambin, E., ... Foley, J. (2009). Planetary boundaries: Exploring the safe operating space for humanity. *Ecology and Society, 14*(2). https://doi.org/10.5751/es-03180-140232

Roy, A. (2020). The pandemic is a portal. *Financial Times*. Retrieved from https://www.ft.com/content/10d8f5e8-74eb-11ea-95fe-fcd274e920ca?fbclid=IwAR3Z7Pa2EeS3r4zS7teY0amvqz BPcDRCuct5Dm8QRtMp1rY_bi0nFUM6WxA

Saulnier, D. D., Blanchet, K., Canila, C., Cobos Muñoz, D., Dal Zennaro, L., de Savigny, D., ... Tediosi, F. (2021). A health systems resilience research agenda: Moving from concept to practice. *BMJ Global Health, 6*(8), e006779. https://doi.org/10.1136/bmjgh-2021-006779%

Shaw, D. (2021). COVID-19 conscience tracing: Mapping the moral distances of coronavirus. *Journal of Medical Ethics*, 1–4. https://doi.org/10.1136/medethics-2021-107326

Shi, L., Chu, E., Anguelovski, I., Aylett, A., Debats, J., Goh, K., ... VanDeveer, S. D. (2016). Roadmap towards justice in urban climate adaptation research. *Nature Climate Change, 6*(2), 131–137. https://doi.org/10.1038/nclimate2841

Southwick, S. M., Bonanno, G. A., Masten, A. S., Panter-Brick, C., & Yehuda, R. (2014). Resilience definitions, theory, and challenges: Interdisciplinary perspectives. *European Journal of Psychotraumatology, 5*. https://doi.org/10.3402/ejpt.v5.25338

Stirling, A. (2010). Keep it complex. *Nature, 468*(7327), 1029–1031. https://doi.org/10.1038/4681029a

Strang, P., Fürst, P., & Schultz, T. (2020). Excess deaths from COVID-19 correlate with age and socio-economic status. A database study in the Stockholm region. *Upsala Journal of Medical Sciences, 125*(4), 297–304. https://doi.org/10.1080/03009734.2020.1828513

Strunz, S. (2012). Is conceptual vagueness an asset? Arguments from philosophy of science applied to the concept of resilience. *Ecological Economics, 76*, 112–118. https://doi.org/10.1016/j.ecolecon.2012.02.012

Sundararaman, T., Muraleedharan, V. R., & Ranjan, A. (2021). Pandemic resilience and health systems preparedness: Lessons from COVID-19 for the twenty-first century. *Journal of Social and Economic Development and Cultural Change*, 1–11.

Tappolet, C. (2013). Evaluative vs. deontic concepts. In *International encyclopedia of ethics*.

Teschl, M., & Comim, F. (2005). Adaptive preferences and capabilities: Some preliminary conceptual explorations. *Review of Social Economy, 63*(2), 229–247. https://doi.org/10.1080/00346760500130374

Thorén, H., & Olsson, L. (2017). Is resilience a normative concept? *Resilience, 6*(2), 112–128. https://doi.org/10.1080/21693293.2017.1406842

Walker, B., Holling, C. S., Carpenter, S. R., & Kinzig, A. P. (2004). Resilience, adaptability and transformability in social-ecological systems. *Ecology and Society, 9*(2). https://doi.org/10.5751/es-00650-090205

Wang, Z., Duan, Y., Jin, Y., & Zheng, Z. J. (2020). Coronavirus disease 2019 (COVID-19) pandemic: How countries should build more resilient health systems for preparedness and response. *Global Health Journal, 4*(4), 139–145. https://doi.org/10.1016/j.glohj.2020.12.001

Wardekker, J. A., de Jong, A., Knoop, J. M., & van der Sluijs, J. P. (2010). Operationalising a resilience approach to adapting an urban delta to uncertain climate changes. *Technological Forecasting and Social Change, 77*(6), 987–998. https://doi.org/10.1016/j.techfore.2009.11.005

Woods, D. D. (2015). Four concepts for resilience and the implications for the future of resilience engineering. *Reliability Engineering and System Safety, 141*, 5–9. https://doi.org/10.1016/j.ress.2015.03.018

Ziervogel, G., Pelling, M., Cartwright, A., Chu, E., Deshpande, T., Harris, L., … Zweig, P. (2017). Inserting rights and justice into urban resilience: A focus on everyday risk. *Environment and Urbanization, 29*(1), 123–138. https://doi.org/10.1177/0956247816686905

Open Access This chapter is licensed under the terms of the Creative Commons Attribution 4.0 International License (http://creativecommons.org/licenses/by/4.0/), which permits use, sharing, adaptation, distribution and reproduction in any medium or format, as long as you give appropriate credit to the original author(s) and the source, provide a link to the Creative Commons license and indicate if changes were made.

The images or other third party material in this chapter are included in the chapter's Creative Commons license, unless indicated otherwise in a credit line to the material. If material is not included in the chapter's Creative Commons license and your intended use is not permitted by statutory regulation or exceeds the permitted use, you will need to obtain permission directly from the copyright holder.